William Libbey, W. W. McDonald

Topographic, Hypsometric and Meteorologic Report

William Libbey, W. W. McDonald

Topographic, Hypsometric and Meteorologic Report

ISBN/EAN: 9783743403048

Manufactured in Europe, USA, Canada, Australia, Japa

Cover: Foto ©berggeist007 / pixelio.de

Manufactured and distributed by brebook publishing software
(www.brebook.com)

William Libbey, W. W. McDonald

Topographic, Hypsometric and Meteorologic Report

*CONTRIBUTIONS FROM THE E. M. MUSEUM OF GEOLOGY AND
ARCHÆOLOGY OF PRINCETON COLLEGE.*

No. 2.

TOPOGRAPHIC, HYPSOMETRIC,

AND

METEOROLOGIC REPORT,

BY

WILLIAM LIBBEY, Jr.,

AND

W. W. McDONALD,

OF THE

Princeton Scientific Expedition,

1877.

SEPTEMBER.

NEW YORK, 1879.

CONTRIBUTIONS FROM THE E. M. MUSEUM OF GEOLOGY AND
ARCHÆOLOGY OF PRINCETON COLLEGE.

No. 2.

TOPOGRAPHIC, HYPSOMETRIC,

AND

METEOROLOGIC REPORT,

BY

WILLIAM LIBBEY, JR.,

AND

W. W. McDONALD,

OF THE

Princeton Scientific Expedition

1877.

SEPTEMBER.

NEW YORK, 1879.

It is with great hesitation and many misgivings that we submit this Report to be printed, being conscious of the fact that errors may have crept into it unknown to us. But, at the oft-repeated request of Dr. Guyot, and the hope that the observations of scientific interest carried on during the Summer of 1877 may be of some value, we have sent it out, knowing that it embodies the results of careful work. That it was done by amateurs who entered into the work from love of it, and who had not the experience of professional men, must be the excuse for any deficiencies which may be found in it. It is with much pleasure, however, that we take this opportunity of thanking Dr. Guyot and Professors Brackett and MacMillan for their kind encouragement and aid, both in the preparation they gave us for work in the field and in the unfolding of the results, after our return home. A number of artotypes enliven the pages of the Report, for the plates of which we are indebted to the courtesy of the photographers of the expedition, Mr. Devereux and Mr. Butler.

WILLIAM LIBBEY, JR.

W. W. McDONALD.

199567

PRINCETON SCIENTIFIC EXPEDITION.

1877.

———

PROF. C. F. BRACKETT, M.D.

PROF. JOSEPH KARGE, PH.D.

W. B. DEVEREUX, '73,
H. R. BUTLER, '76, } *Photography.*

R. H. LYNDE, '77,
H. F. OSBORN, '77,
J. POTTER, '77, } *Palæontology.*
W. B. SCOTT, '77,
F. SPEIR, JR., '77,

A. J. McCOSH, '77,
C. G. GREENE, '77, } *Botany.*

J. S. ELY, '77,
F. E. PARKER, '77, } *Mineralogy.*
W. DULLES, '78,

W. F. DUNNING, '78,
D. F. McPHERSON, '77,
W. D. VAN DYKE, '78, } *Natural History.*
J. A. STEWART, JR., '79,

W. LIBBEY, JR., '77,
W. W. McDONALD, '78, } *Topography, Meteorology, Hypsometry.*

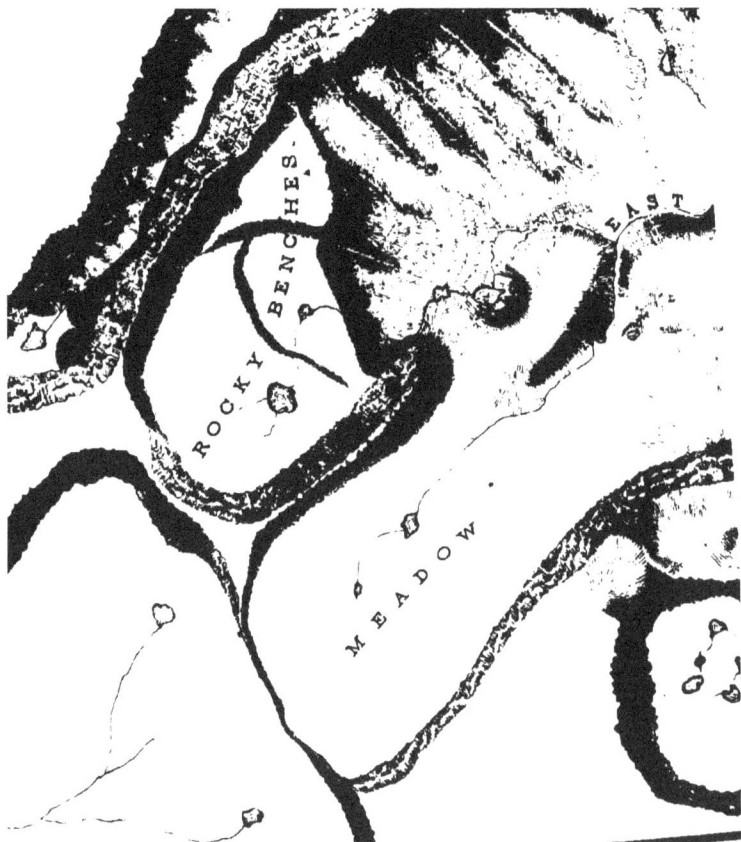

WEST BRANCH

BRANCH OF SMITH'S FORK.

HENRY'S FORK.

Princeton Scientific Expedition.
1877.
MAP OF A
TOPOGRAPHICAL RECONNAISSANCE
OF THE
HEAD WATERS
OF
SMITH'S FORKS
IN THE
UINTAH MOUNTAINS
UTAH.

NOTE. All spaces left blank, unless otherwise designated, are unexplored.

William Libbey Jr 77
W.W.McDonald 78   Topographers.

SCALE OF FEET.

To the Honorable the President and Board of Trustees
of the College of New Jersey.

Gentlemen : I have the honor to submit herewith the report of the two topographers in charge of the topography and meteorology of the college scientific expedition of 1877.

The map of a portion of the Uintah Mountains, which is exclusively their work, has been reproduced by the artotype process ; and, together with the numerous altitudes carefully determined, is believed to be a valuable contribution to the knowledge of the character of that chain of mountains, as yet so little known. The artotype plates are copies of but a few samples selected from the large collection of excellent photographs, taken by the photographers of the expedition. The appendix, containing a diary of the party, is given to show how faithfully the time allotted has been employed in furthering the object of the expedition. It may serve as a specimen for all the divisions, which would show a similar record.

It is a great pleasure for me to be able to state that a further examination of the palæontological results of both the expeditions of 1877 and 1878 have much surpassed my expectations. A second palæontological report is almost ready for publication. The materials for a third, containing a large number of new species of fossil plants, are nearly prepared.

It is a source of great gratification that these expeditions, while accomplishing their first object, the encouragement of scientific studies in our college, have been at the same time so fruitful in results for the advancement of science. This may justify the hope entertained that this doubly useful complement of our regular course of studies will remain a permanent feature of our College.

Very respectfully submitted,

A. GUYOT,

Director of E. M. Museum.

Princeton, N. J., September, 1879.

PRINCETON, N. J., June 1, 1879.

SIR : We transmit herewith the reports upon Topography, Meteorology, and Hypsometry, as prepared by Messrs. William Libbey, Jr., and W. W. McDonald, of the Scientific Expedition which went out in the summer of 1877 under our charge.

Together with the reports are three maps : the first containing the results of a topographical reconnaissance in the valley of Smith's Fork, in the Uintah Mountains of Utah ; the second containing the triangulation of the same ; the third showing the relation between Fort Bridger and the valley of Smith's Fork.

The topographical report is divided into two parts, one devoted to the work in Colorado, the other to that in Utah.

The meteorological report is confined wholly to Utah, and contains a meteorological record for part of the time spent there.

The hypsometric report is also divided into a report of the work in Colorado, then of that in Utah, and these are followed by a discussion of the barometric work.   Yours, etc.,

C. F. BRACKETT.

JOSEPH KARGÉ.

DR. A. GUYOT,

Director of the E. M. Museum.

PRINCETON, N. J., Jan. 1, 1879.

SIRS: We present herewith our joint report on Topography, Meteorology, and Hypsometry, embodying the results of our work in those departments during the summer of 1877. Accompanying the report is a map, prepared by us, giving an idea of the topographical features of the upper portion of the Smith's Fork Valley in the Uintah Mountains. This valley was chosen as a characteristic type of the valleys of this transverse mountain chain and studied as such with some care.

Leaving Princeton June 21st in the evening, we reached Denver on the evening of the 25th. Here we spent a few days in purchasing such things as we needed to complete the equipment brought from the East. The party also bought horses, mule teams, and wagons here for their proposed trip in Colorado. From this time to July 21st we spent in Colorado; measuring Pike's Peak, Mts. Lincoln, Bross, Quandary, Silverheels, Princeton, Evans and Gray; also, determining several bases. We then started with General Kargé and the Palæontologists for Fort Bridger. Our trip was made with safety and pleasure, notwithstanding a delay at Cheyenne. We fitted out at the Fort for the mountains, where we worked at the topography of the valley mentioned above until August 24th, and then commenced our return to Fort Bridger, aiming, on our return, to get a general idea of the region lying between the Fort and the mountains. The result of this reconnaissance is embodied in another map, also accompanying this report, which may serve as a hint to further and more accurate surveys of the same region. The meteorological as well as the topographical report is confined mostly to this part of Utah. The only work of any scientific value done in Colorado was hypsometric, on account of our hurried and long marches through that State. Being constrained to keep track of our party, we depended on gaining some miles in advance for an opportunity of doing our work. The work in Utah was undertaken in a more systematic way. A base line was accurately measured, and its height carefully ascertained; from either end of this our triangles stretched up the sides of the valley of Smith's Fork, and these were bound together by a network, in the measurements of which we have some confidence, not only on account of the accuracy of our little transit, but from the knowledge that the work was done with great care. We worked up one side of the valley first, and then down the other, our camp being about the centre of the large amphitheatre. The last triangle closed within half a minute of arc, which we considered excellent, from the fact of the roughness of the ground traversed and the many disadvantages under which we labored. The peculiarities of our work will be brought out fully in the body of the report. Our experiences with electricity, and the elements in general, may not have been exceptional, but they were certainly novel to us, and their faithful recital may prove a warning to future explorers in the same region. Upwards of thirty barometric measurements were made of the mountains on either side of the valley. Our series of barometric heights commenced at Carter Station on the Union Pacific Railroad, and extended from there to Fort Bridger; thence in succession to the Herd House or cattle yards of Judge Carter ; to the St. Louis Mills ; to Gilbert's Meadows ; and to our other two camps, the one near the head-waters of the west branch of Smith's Fork, and the other at the centre of the valley of the east branch, or main Smith's Fork. Our barometers were carefully compared before and after each trip, or day's work, and the station barometer was always carried as near to the base of the mountain as was possible. A north and south line was preserved as much as possible in the relative situation of the instruments. In ascending the side of a mountain we were often enabled by means of our levels to know when we were on a line with the top of some previously measured mountain on the other side of the valley, and by measuring this point also we were often able to obtain a second measurement, which was in general very satisfactory in its comparison with the former one. Towards the latter part of our stay the rainy season of the fall came on, and heavy rains, with storms of snow and hail on the mountains, generally lasting from 10 A.M. to 4 P.M., drove us from them. Bidding them a reluctant farewell, we returned to the Fort, making measurements by barometer, sketching profiles, and making sections, with which, and with the aid of some other maps at the Fort, we finished the map of the country between the Fort and the mountains. The hypsometric and topographic work are combined in the main map accompanying the report, which has been prepared with great care. By comparing our map with that of the government survey of the fortieth parallel, it will be readily seen that it differs in some particulars from the map of the same valley given by that survey. It cannot be expected from a general survey, such as was intended in that work, that it should contain all the details of a special survey such as ours. To a person merely passing through the eastern part of the valley it presents many of the aspects given on the above-mentioned map ; but the appearances in that valley are very deceptive, and a closer examination reveals the features given in our map. After the regular reports we have given an itinerary of the whole trip, believing it will be interesting to those who wish to follow us in our daily work in the West, as it combines, in a condensed shape, the diaries of both of us. Our thanks are due to Colonel Flint, commandant of the Fort ; to Lieutenant-Adjutant True, Lieutenant Scott, commissary ; Judge W. A. Carter, Dr. J. Van A. Carter, and Mr. R. H. Hamilton, for many courtesies and great aid in the furtherance of our work. We shall long remember with pleasure our leisure hours passed among the pleasant people of that little island in the desert, where we met so many kind friends. And the warm reception, with the kind-hearted farewells, will always raise happy memories in that division of our scientific expedition. Leaving a few days later (September 6th) we joined the rest of the party at Cheyenne : then with only a stop at Chicago over Sunday, we set our faces eastward, with little to regret and much to be happy over.

WILLIAM LIBBEY, JR.
W. W. McDONALD.

C. F. BRACKETT, M.D.
JOSEPH KARGÉ, Ph.D.

INSTRUMENTS.

The instruments of the expedition were the property of a member of the party, and were as follows :

TRANSIT.—The transit was made by Messrs. Heller & Brightley, of Philadelphia, and is a *fac-simile* in miniature of their "complete engineer's transit." It is one of the instruments exhibited by this firm at the Centennial Exhibition in Philadelphia, where it took a prize. It has long compound centres, the vernier plate centre being 2½ inches long. The horizontal limb is read, by two double opposite verniers, to minutes of arc. These verniers are placed outside of the compass-box, and their openings are protected by glass windows. The vertical arc also reads to minutes. There is a 2½-inch magnetic needle, and its ring is divided into half degrees. The telescope is 6¾ inches long, with object-glass $\frac{16}{8}$ inch in aperture ; it magnifies 15 diameters, and shows objects erect, not inverted. A sensitive level 3 inches long is attached to the telescope for assistance in accurate levelling work. The telescope has a slow motion for focussing the cross wires, and is provided with adjustable stadia wires for measuring purposes, thus dispensing with chaining and taping to a considerable extent. The plates and vertical circle are provided with clamp and tangent screw movements, and the clamps on the axis of the telescope are arranged with sighting slits and indexes, so as to answer also for right angle sights. All the working parts of the needle-lifter, clamp and tangent screw movements are concealed between the plates, making the instrument more compact and less liable to injury. The height of the instrument from the tripod legs is 7¼ inches. The extreme diameter of the plates is 4¼ inches. The diameter of the horizontal plate at the point where the verniers and graduation meet is 3⅝ inches. The tripod is furnished with an adjustable head for precise plumbing over a centre. The weight of the instrument exclusive of tripod head and legs is 5 lbs. The weight of the tripod is 3 lbs. The instrument, with its various additional parts (such as sun-tube, screw-drivers, magnifier, etc.), is packed in a box 8 inches long, 5 inches wide, and 7½ inches deep. This box is arranged with straps to allow of its being carried over the shoulder. The constant to be added to every recorded stadia reading of this instrument was obtained as follows : Object focal length, 4⅞ inches ; object-glass to centre of instrument, 3⅜ inches. Hence, if 8¼ inches be added to the stadia measurement, the measurement will then be to the centre of the instrument or plummet. We found this instrument reliable, its graduation very good, and the power of its telescope remarkable.

SEXTANT.—The sextant was one of Stackpole's best. It reads to 10' of arc, and was provided with all the accessories usually accompanying such an instrument. The pocket sextant was made by James Green, of New York, and read to minutes of arc. Both these instruments did good service.

BAROMETERS.—The two mercurial and two aneroid barometers used were made by James Green, of New York, especially for this work, and were the best instruments he could produce. The two mercurial barometers had double verniers reading to ·002 of an inch, and were capable of recording 18,000 feet elevation. They were each packed first in a wooden case, which was then placed in a leathern sheath, and thus thoroughly prepared for rough usage and mountain work. They worked admirably together, as was shown by a long series of comparisons before and after leaving Princeton. A comparison was also made before and after each day's work, and the extreme variation was never found to exceed ·004 of an inch. We carried with these instruments a case of repairing articles, including bottles of mercury, leather, thread, etc., also six extra tubes in a box, so that in case of an accident we would not be left hopelessly in the lurch. But such was the quality of the workmanship displayed in these instruments that notwithstanding the exceedingly rough usage they received, not a repair was necessary, and they were brought home in as good working condition as when they started out. Of the two aneroids, one read on its face to 20,000 feet, with 50-feet divisions. It was also divided to 14 inches, again divided to hundredths. The other read to 15,000 feet, with 50-feet divisions, and was also divided to 14 inches, and hundredths. These two instruments were selected from six first-class ones as the best of the lot. They were tested by trials under the air-pump in the physical laboratory at Princeton, and worked well, though they have not done so well since their return.

LEVELLING-ROD.—The levelling-rod was made by Keuffel & Esser, of New York. It was an engineer's telescopic rod, and read to thousandths of a foot.

LEVELS.—The two levels were the ordinary pocket-levels, with level-tube on the inside of the tube, and an inclined mirror at the bottom of the same to reflect the image of the bubble. They were made by Mr. Green, of New York.

PRISMATIC COMPASS.—This instrument was also made by Mr. Green, and of the same pattern as is used by the United States Engineer Corps.

COMPASSES AND CLINOMETER.—These were made by Mr. Green. There were three small compasses for pocket use in brass cases. The clinometer was a 3-inch compass, mounted upon a square piece of brass, one edge of which served as a base for the clinometer index. This compass was also provided with sighting slits.

TELESCOPES.--These were made by Bardou fils & Cie., of Paris, having glasses of 1¼ and 2 inches diameter. Both of them were pieces of fine workmanship, and were of great service on account of their clear definition and high powers.

THERMOMETERS.—These were graded according to Fahrenheit and made by Mr. Green. Each had a wooden case of its own. One of them was used for a wet bulb thermometer in meteorological observations.

TOPOGRAPHICAL REPORT.

I. COLORADO.

I UTAH.

Gateway, in the "Garden of the Gods."

TOPOGRAPHICAL REPORT.

I.—COLORADO.

OUR work in Colorado was almost entirely hypsometric; but as we rode a great many miles on horseback through the most beautiful and picturesque parts of that State, and in that way obtained an idea of the topography of the country through which our road lay, we shall give some account of it. We were materially aided in this by the Government reports of Dr. Hayden, which have left very little to be done in that region in any of the departments of scientific investigation. In this report, the plan of which has been to gather together all the information possible on each of the special points of interest that we visited, we shall aim to do little more than glance at those features which were observed by us, and therefore our description may seem as hurried as our trip necessarily was. The pictures we have inserted will almost explain themselves; but a few words will sometimes be quite proper to guide those who have never seen the places referred to. The description will follow the course of our trip, and on this account may seem somewhat disjointed; but our excuse is that there was no other course left open for us; any one, therefore, who expects a complete topographic report of this part of our work will be disappointed. All we can hope is to give such information on the mountains we measured and the places we visited as might be instructive to any one interested in the subject. With these few words of introduction we will proceed first to the description of the Front Range as seen from Denver, and then to a nearer view of its structure and of that of the land to which it forms a border.

As one approaches the Rocky Mountains from the prairies, the Front Range is the first object which greets the eyes; and it is rather a pleasant sight after the long journey over the plains. This feeling is also mixed with surprise at the sudden change from plain to mountain, the cause of which has been a puzzle to many a traveller. The abrupt change from plain to mountain is not more sudden, however, than the change in the geological composition of the two systems under consideration. The mountains rising along the border line present in their rough and precipitous faces the strong, bold outlines produced by the metamorphic rocks, the granite, gneiss and the shists; while the sedimentary rocks at their bases, the limestones, shales, slates, clays and chiefly sandstones, present a very different appearance.

This border line extends in a nearly north and south line about twenty miles west of Denver, and is admired for the beauty of its outlines. The rapidity of its slopes (a marked feature of the chain) afford a grand opportunity for the display of the effects of erosion, the summit of the range being in many cases scarcely ten miles from the edge of the plains. Any one looking at the map of Colorado will also be struck by the beautiful arrangement of these chains *en echelon*.

Geologists tell us that this range was first elevated from the sea, and then that all these sedimentary rocks were deposited against their base. These sedimentary rocks at their point of contact with the granites are not horizontal, but have been turned up on end by the force exerted by the granite mass, against which they were deposited, when it lifted them to their present positions. Their slope to the east from this point is not more than 60° at any place. They form the basis of the great plain, and were once the bed of the sea which covered them, and which on receding left those thousands of feet of sediment to be carved out and carried off by the ice and water, which have left such wonderful traces of their power in this western country.

The first feature of the lowlands that strikes the eye is the low series of hills of a very even line of elevation, forming a sort of horizon or belt near the foot of the mountains, cut at intervals by the streams which descend from the mountains. The ends of these hills that overlook the streams being afterward rounded, give each section a long, gently curved line at their upper surface, which has earned for them the rather unfortunate name of "hogbacks." They seem to be a very natural boundary line between the two geological systems here brought together.

It is now believed that the sedimentary rocks, the ends of which are now exposed to our view along the eastern side of this valley, must have extended very much farther into the mountains, and that after the elevation of their mass they were partly removed by erosion. The thickness of these layers is given by Dr. Hayden's surveys at 7000 feet.

From the consideration of these general features let us turn our attention to the various phenomena as seen when travelling along between the granite and the sedimentary rocks. No one can obtain an adequate idea of the wonders of this famous strip of country until he has seen it for himself, and it would be hopeless for us to attempt a description that would fully convey the impression produced by these master-pieces of the work of erosion. We shall therefore be content with the scientific part of the description, leaving the imagination to fill up the blanks.

The Front Range is really broken but once below Denver by a deep cañon, where the Platte rushes from the South Park through this narrow path to the plains. Below this point, on the road to the south, we cross the low divide which separates the Platte from Plum Creek, and soon descend into the valley, whose curiously carved monuments, of which we will speak below, make it so attractive.

About half-way between Denver and the Garden of the gods we find the divide which separates the waters of the Platte from those of the Arkansas. This slight elevation of a little over 1000 feet above Denver controls the flow of the tributaries of those streams until they reach the open plains to the north and south of this point, when they take the direction of the long gentle slope to the east. To the south of this divide these monuments increase in number until the Garden of the gods is reached, which owes its name to the grand display of these singular objects at that place.

These monuments seem to have been formed in several ways, principally, however, by erosion; these isolated parts have been left because they were either of harder material than that surrounding them, or more able to resist meteoric influences, or because of their position with regard to the currents. They seem to be an aggregation of quartz grains and pebbles loosely held together in a nearly circular column, which tapers from the thick base towards the top. This shaft is surmounted by a cap of rust-colored sandstone, which owes its greater size to the oxide of iron that forms a cement, binding the grains together. This sandstone being on that account a much harder compound than the layers below it has resisted erosion to a greater degree. The smaller monuments here described vary in height from 10 to 20 feet. There remain, however, some other forms which must not be omitted in this description, and those are the castellated forms of the larger table buttes or "mesas." These massive objects are from 100 to 250 feet in height, and are sometimes capped with a layer of purple porphyritic basalt. They rise from the beautiful green meadows, and their almost perpendicular sides give them an appearance which is particularly impressive just at sunset.

There seems to be but little doubt that this line of upturned strata exists throughout the whole length of the Front Range, though hidden in some parts of its length. It is separated from the Front Range by a valley, the portion of which that lies fifty miles to the north of Colorado Springs is the most picturesque, and will always retain its celebrity, on account of the striking forms of erosion just described.

Pike's Peak is situated just west from Manitou and the Garden of the gods. It is a mountain composed of fine-grained reddish granite, which is believed to be metamorphic. The appearance of some fragments of sedimentary rock at its foot gives rise to the supposition that in Silurian times there was a bay at its base, this being the point farthest west at which these rocks are found in this range. It is also the point where the summit of the range approaches nearest to the plains ; the upheaving force seems here to have been expended in elevating the older layers above the sea, making them form part of the shore-line and driving the sea more to the east ; whereas to the north of this the force being more distributed caused a more general elevation. The same is supposed to be true with regard to the Mt. Evans group at

Single Butte, in "Garden of the Gods."

the other extremity of this line, both of these mountains seeming to be centres of action.

The Ute Pass lies to the north of Pike's Peak, and is a very beautiful and picturesque pass, cut out of the granite along the course of the " Fontaine qui bouille," and is the road to South Park, which comes to our notice next.

South Park is a large area of depression, of the general shape of an ellipse, having a diameter of 45 miles from northwest to southeast, and of 35 miles from northeast to southwest. Its area is about 1200 square miles. It is not a perfectly level plain, but has a general slope from the northwest, almost to the base of the Front Range, where lies the bed of the Platte. This accounts for the curious drainage system of the Park, as a moment's glance at the map will show. The rivers flow down the long slope of the western side till they reach the short counterslope of the eastern side, and then turning almost at a right angle flow to the northeast. The general elevation of the Park is about 8000 feet, though it reaches 10,000 feet in the western part, and goes down to 7000 where the Platte cañon pierces the Front Range. It is surrounded by mountains, the bases of which bear the marks of its having been at one time the bed of a lake, not only from the form of the basin, but mainly from the sedimentary rocks found deposited there. The surface of the plain is quite irregular, being cut up by numerous low ridges, which are generally parallel to each other. In the central portion there are also cross-ridges of trachyte, showing the existence of some extended fissure from which this igneous overflow took place. Otherwise the sandstones of the Park are very little disturbed.

Before describing either the Park Range or the Sawatch, let us look at the valley of the Arkansas, which lies between them. It starts at the Tennessee Pass from a basin of granite, and has been formed partly by fissure and partly by erosion. This valley varies in width from 8 to 10 miles, and is cut down a hundred miles through the mountains to Poncho Pass. It affords one of the finest fields for the study of glacial action in the West. Dr. Hayden thinks * that the

* U. S. Geog. and Geol. Survey, Hayden, 1873, p. 39.

2

Park Range is a member of a great anticlinal, of which the Sawatch is the nucleus or central axis. This valley coming between these ranges was thus first opened by fissure, and then the enormous work of erosion commenced. At first the valley must have had a huge glacier running from north to south through its length, as there are still traces of such a glacier in the markings on the sides of the mountains and in the drift matter on its slopes. Then part of this valley formed the bed of a large lake, as is shown by the deposits in the bottoms where they are exposed. This lake occupied the lower half of the valley, and when it was drained off through the opening now traversed by the Arkansas River the heavy and coarse material at the upper end and the finer drift matter at the lower end were exposed. The valley has many rounded oblong hills, which are covered by débris, and range in height from 500 to 700 feet. After this first large glacier came others, which might be called secondary, and occupied the beds of the present tributary streams of the Arkansas. Each one of these is marked by large moraines, and where exposed, the glaciation is magnificent. The masses of rock which have been transported by these agencies are incredibly large, often reaching 100 feet in diameter. The largest of these valleys is that of Lake Creek. This whole valley must have been occupied by a glacier of from 1000 to 1500 feet in depth. The terminal moraines of this glacier are remarkable for their size. Everywhere the traveller is hindered in his journey by mounds, ridges, basins and boulders, the latter often from 20 to 50 feet in diameter. Worn rocks are also exposed, showing the effect of ice on their surfaces. The upper part of the valley has two lakes in it, whose basins were doubtless formed by the glaciers. These lakes are about 350 yards apart and connected by a small stream. The lower of these lakes is the largest, being 2½ miles in length and 1½ in breadth, its greatest depth being 75 feet. The upper lake is 1 mile in length and ½ mile in breadth, having a depth of 79 feet.

On either side of this valley are the Sawatch and Park Ranges, the origin of which we have already indicated. The great mass of the Sawatch Range is a granite nucleus, with now and then a dike of some foreign material. There is one

very remarkable feature about this range, and that is the superficial covering of earth which reaches clear to the summits of many of the peaks. About the middle portion of the range are situated the three prominent peaks of Princeton, Yale, and Harvard; respectively, 14,208, 14,187, 14,375 feet above the level of the sea.

Its companion, the Park Range, is very irregular in height, being much cut to pieces by erosion. One of its marked features are the trachytic beds which are interstratified among its other rocks; these layers have, in places, a thickness of 1000 feet. Another peculiarity is the presence of the large amphitheatres at the head of the streams which arise on its eastern slope. It is very precipitous on the western side, its gentle inclination to the east giving rise to the supposition of its being an anticlinal.

Little remains for us to do now but to glance at that knot of mountains which centre in Mt. Gray, forming a connecting link between the Front and Park Ranges, two of whose peaks we measured: Mts. Evans and Silverheels. As the main crest of the Front Range approaches Mt. Torrey, it suddenly rises from its usual height of about 12,000 feet to the magnificent elevation of over 14,350 feet. From Mts. Gray and Torrey a very imposing and high ridge runs out to the east, which reaches its climax in Mt. Evans. From Mt. Gray the Range to near Mt. Lincoln is much higher than in any other part. We crossed the Argentine Pass at 13,100 feet, and the Hoosier Pass at 12,364 feet. Mt. Silverheels is on this ridge just before it reaches Hoosier Pass, and is also the centre of another group of mountains. It might be said that the main ridge reaches here its maximum height, for a pass at this point is 13,650 feet. Near Mt Lincoln the main ridge meets the Park Range at right angles, and we shall leave it at this point to follow the Park Range far enough to mention the three mountains we measured on its crest. The Park Range has four mountains at this point, all within a short distance of each other, which are above 14,000 feet; of these we measured three: Mts. Quandary, Lincoln and Bross. It is also noted for its high passes, the number of its peaks, and the amphitheatres at the heads of the several valleys. This brings us to the end of our work in Colorado.

PICTURES.

The pictures we have been able to give are as follows:

THE GATEWAY IN THE "GARDEN OF THE GODS." The two large remnants of a perpendicular stratum of red sandstone which was cut in two by erosion form this gateway. In the foreground is seen a stratum of white sandstone, also on its edge. In the distance Pike's Peak is dimly outlined.

A SINGLE BUTTE. This has already been described, and will be readily recognized.

GLEN EYRIE. We give two views; one of the "needle rocks," as a remarkable example of the work of erosion, and the other of the edge of one of these upturned ledges which gives a fine idea of its structure.

MT. PRINCETON. This mountain is in the Sawatch Range. It is noted for its symmetry. The broad valley of the Arkansas is in the foreground.

THE WESTERN TWIN LAKE. This is the smaller of the two lakes. Two unnamed peaks of the Sawatch chain are shown in the background.

WESTERN SLOPE OF THE CONTINENTAL DIVIDE. This view is in the Twin Lake region.

. BLUE RIVER AND MOUNTAINS. This view is given as a specimen of a deep and wooded valley. The slope of the mountains, it will be noticed, is very steep. The height of the ridge above the valley is 3000 feet.

MT. LINCOLN. The view is taken from Hoosier Pass, and embraces Montgomery Gulch on the right hand side of the picture. The trachytic dikes spoken of in the report are well brought out on the side of the mountain.

MT. GRAY. The view is taken from the Argentine Pass, and shows Mt. Torrey on the right.

Glen   Eyrie
No. 1 1.

II.—UTAH.

Concerning the topography of the Uintah Range, how-ever, we will go a little more into detail; especially in regard to the valley, which we endeavored to study with more care than any other part. A careful comparison of what may be said here with the map before one's eyes, will give an ade-quate idea of the many beauties, as well as the remarkable structural features, of the valley. We regret very much that we are not able to give any pictures of this strange place. Prof. O. C. Marsh says[*] of the Green River Basin, in the March number of the *American Journal of Science and Art* for 1871, p. 197 :

"While descending the northern slope of the mountains toward the great Tertiary Basin of the Green River, which lay in the distance 2000 feet below us, we passed over a high ridge, from the summit of which appeared one of the most striking and instructive views of geological structure to be seen in any country. Sweeping in gentle curves around the base of the mountains from where we stood, many miles to the northward was a descending series of con-centric, wavelike ridges, formed of the upturned edges of differently colored strata, which dipped successively away from the Uintahs, those nearest to us 40° or more, those at a distance seeming but little,—altogether a scene never to be forgotten. Apparently, we had before us a geological series from the Palæ-ozoic to the Tertiary."

Fort Bridger is situated in this valley, about fifty miles to the north of the point from which, we suppose, this view was taken, and we can do no better than to quote Prof. Marsh once more, when he says :

"Fort Bridger is situated at the northern base of the Uintah Mountains, about 7000 feet above the sea. The surrounding plain is part of a great basin of denudation, washed out of light clays and soft sandstones of the Tertiary age ; the deposits in one of the great fresh-water lakes that replaced the Creta-ceous sea, from which the mass of the Rocky Mountains emerged. Remnants of the strata removed may be seen at various points around ; some in the shape of flat, isolated buttes, and others forming benches resting horizontally against the sides of the mountains. These fragments serve to show the great original thickness of this lake deposit, which cannot have been less than 1500 feet, and may have been much greater."

The Uintah Range is a system which leaves the Wahsatch Mountains in about latitude 40° N. with an elevation of nearly

[*] See also Dr. Jos. Leidy, Geol. Survey of the Ter., 1872, p. 651.

12,000 feet, and runs east throughout its whole length; at
first, a little north of east and then with a single bend going
southeast. It seems to reach its maximum in Gilbert's Peak,
which cannot be far from 14,000 feet of elevation. From this
point it descends till it terminates near the Green River.
The range forms an immense ridge, with an almost level, but
badly broken up belt on top, which has an elevation of about
6000 feet above the table-land on the north, and apparently
about 7000 feet above the similar table-land on the south.
The valleys are numerous, deep, and narrow, except at the
head, and extend quite to the summit. Between these valleys on the north are the transverse and generally unbroken
ridges by means of which the summit can be reached by a
gradual ascent. On the south these transverse ridges seem
to be very much broken toward the summit. The opposite
valleys start from nearly the same point, and have between
them only a thin precipitous wall, which erosion is slowly re-
moving. Gilbert's Pass at the head of Smith's Fork is such
a point, where the wall has been sufficiently worn away to
admit of the construction of a road across it. There are few
if any other such places at present.*

The sides of the ridges which bound these valleys are
generally masses of boulders and débris *in situ*, making their
ascent almost an impossibility, and travelling a very difficult
operation. The country to the north from the end of these
ridges descends in a series of terraced plateaus of soft earth,
the slopes of the terraces being generally as steep as the
earth will stand.

The whole top of this elevated belt of country is very badly
broken up by cracks in many places, affording a fine chance
for the study of dynamical geology. It seems to have been
the favorite battle-ground of the elements, some of the peaks
having-been literally crushed into a mass of huge boulders.

Our topographic survey in this valley has been carefully
worked out and embodied in the principal map which accom-
panies this report. The diary of the trip will give the daily
occurrences, and for this reason it will only be necessary
here to indicate the route taken and the bases which were

---

* See Geol. Survey of 40th Parallel vol. ii. p. 257.

Mt. Princeton.

measured. Starting out from Carter's station, and using the height given by the Union Pacific Railroad (6530 feet), we ascertained the height of Fort Bridger to be 6714·8 feet. Then by a series of half-day corresponding observations and comparisons we moved on up into the mountains, heading toward Gilbert's Meadow, which, although we found it a good camping place, was not a point for active work, and hence we had to move on to a camp at the head of the west branch of Smith's Fork. We remained there till the place of our day's work became too far off for us to reach it comfortably. We then moved to the centre of the East Branch of Smith's Fork's valley, upon one side of which we had been working. From this base we operated most successfully, till we were forced to make our way back on account of the storms which occurred daily. We went back slowly, verifying our work, and returned to Fort Bridger to find our circle of elevations come to a close within two feet. Although the result may be an accident, still this speaks for itself. The barometric bases measured were as follows:

Fort Bridger	6714·8
St. Louis Mills	8557·6
Gilbert's Meadow	9733·2
West Branch of Smith's Fork	10626·0
East " " " "	10555·0
Steel's Mills	9357·0
Herd House	7085·0

The points measured from these bases were as follows:

No. I	10796·4
" IV	10814·0
" V	11910·0
" VI	11605·5
" VII	11397·0
" VIII	12804·8
" IX	12799·5
" X	13052·0
" XI	12804·0
" XII	12622·9
" XIII	12688·2
" XIV	12245·6
" XV	13069·7
" XVI	12401·9
" XVII	12528·8
Timber Line	11144·4

Geologically considered, the Uintah Mountains are the result of an upheaval which followed closely the Carboniferous
Epoch.   In this upheaval a great belt of country from 25 to
30 miles wide was raised to its extreme elevation, without
materially disturbing the inclination of the strata ; it being
possible that the present dip, which increases from 3° at the
summit to 6° at the south, is the same that existed before the
mountains were lifted.   Upon both edges of this belt the
strata are broken axially, dipping on the north side 38° to
the north, 12° 12' west, from the true meridian, and on the
south about 43° to the south, 12° 12' east, in the longitude of
Gilbert's Peak.   The main anticlinal axis is on the north edge
and in the same longitude bears north 77° 48' east from the
true meridian.

The uplifted beds, as displayed by the lateral erosion, are
almost entirely composed of a brownish red—rarely gray—
sandstone of the subcarboniferous epoch, scarcely fossilifer-
ous, which is metamorphosed into quartzite along the anticlinal cracks.   They are of very great thickness, from 2500
to 3000 feet being visible near the summit.

Along the main anticlinal axis, and especially across the
uplifted belt previously alluded to, are numerous lateral
cracks that have been the starting-points of an amount of erosion that seems wonderful.   They have been worn into enormous amphitheatrical basins from 2000 to 3000 feet deep,
from two to three miles wide at the head, which gradually
widen till they reach a point some four or five miles down
the valley, and then again they come together in a cañon
or precipitous valley, which continues for some eight or ten
miles below.   Those on the north are generally worn quite
across the axis into the high belt beyond.

The large basins, previously described, range along either
side of the summit line, and thus have a direction consid-
erably oblique to that of the prevailing winds, and, in con-
sequence, the amount of snow which drifts into them during the winter must be enormous.   At such an altitude this
snow melts comparatively slowly, and furnishes an almost
continuous supply of water to the very numerous mountain
streams, much of it having been first caught in the multitude
of small lakes which are sprinkled along in the basins.   On

COUNTRY BETWEEN
FORT BRIDGER
AND THE
UINTAH MOUNTAINS

SCALE OF FEET.

the north side there is a great deal of marshy land in these basins; but on the south the water, having to make about 1000 feet more descent, runs off too rapidly to leave any marsh. This circumstance, combined with the greater heat of the southern exposure, makes many of the smaller streams on that side run dry early in the season. The streams heading in these mountains are characterized by main features which are common to all. They are generally quite small, and, taking their rise in the summit basins, they flow on the north side, through narrow gorges with steep, heavily-wooded slopes, until they emerge from the foot-hills in the terraced plateaus which abut against the base of the range. Above the plains their water is almost pure, from the fact of its flowing only over sandstone. This quality, together with its great coolness, makes the head-waters of these streams the favorite abode of a very superior kind of mountain trout. Below the upper limit of tree growth, which is at about an altitude of 11,000 feet, the mountain ridges are covered with a dense forest, very much interspersed with small open meadows, with their grass often spreading itself for a considerable distance into the timber.

The valley bottoms are rich and pretty generally covered with grass; willow and cotton-wood trees grow near the water, with groves of pine on the mountain slopes. Everywhere, even to the borders of the snow-banks almost at the mountain heights, there is a luxuriant display of flowers, and grasses thrive along the top of the ridges, above the limit of tree growth and quite up to the summit of the range.

The soil of the plateaus between the streams is almost exclusively yellow marl, mingled with, and overlain in many places by broad lines of drift from the mountains. It is quite rich, and, under different climatic influences, fit to sustain vegetation. At present only sage-brush, grease-wood, and bunch-grass grow upon it.

The whole of the mountain slopes, from the edge of the terraced plateaus to an altitude of about 11,000 feet, are covered with extensive forests of white and yellow pine, spruce, red cedar, hemlock, and aspen. Nearly all of these woods were of young growth; thus giving evidence that extensive fires have raged here at intervals over a long period of time.

We commenced our triangulation in Smith's Fork Valley by establishing a base-line on the top of the long ridge which runs between the east and west branches of the river. This line was measured very carefully. A difference of $\frac{3}{10}$ of an inch only was found between the several measurements. From the ends of this line our triangles stretched up the side upon which our camp was situated, and across to the other side. It will be seen from the triangulation map given in the report that some' parts of the valley figured in the main map were not reached by us; these we were enabled to fill in from sketches which we took from several points, as was our custom whenever it did not rain too hard. The height of No. 11 was ascertained from the top of No. 12, which was seen to be nearly on a level with it, so nearly that the small fraction of a foot given fully covers the difference. Where it was possible, we of course measured all the angles of our triangles, but as we never reached the summit of No. 11, on account of its precipitous sides, we could not make the measurement of the angles centering there. It formed, however, a fine triangulation point, as it had a sharp, spire-like head, which was situated just at the summit of cliffs rising perpendicularly almost 2000 feet from the valley below. We experienced little difficulty in finding our monuments, as the air was remarkably clear when we could work, and probably this accounts for the accuracy of our triangles; they closed within half a minute of arc, and, as our transit only read to minutes, we thought they would answer our present purpose. The measurement of the triangles was always repeated three times and the mean of the three readings taken. Compass directions of the lines were also taken when the electricity permitted. We had two camps in our work in this region; one in the little rounded valley at the head of the west branch of Smith's Fork, on the left-hand bank; the other on the little hill directly at the base of No. 11, on the lower side of the hill, about twenty yards from the right-hand head of the east branch of the same river, at a point a little above the place where it joins the other head. The topography of this valley we have found very hard to represent on paper by means of topographical emblems on account of several features which seem to be entirely peculiar to it—we refer

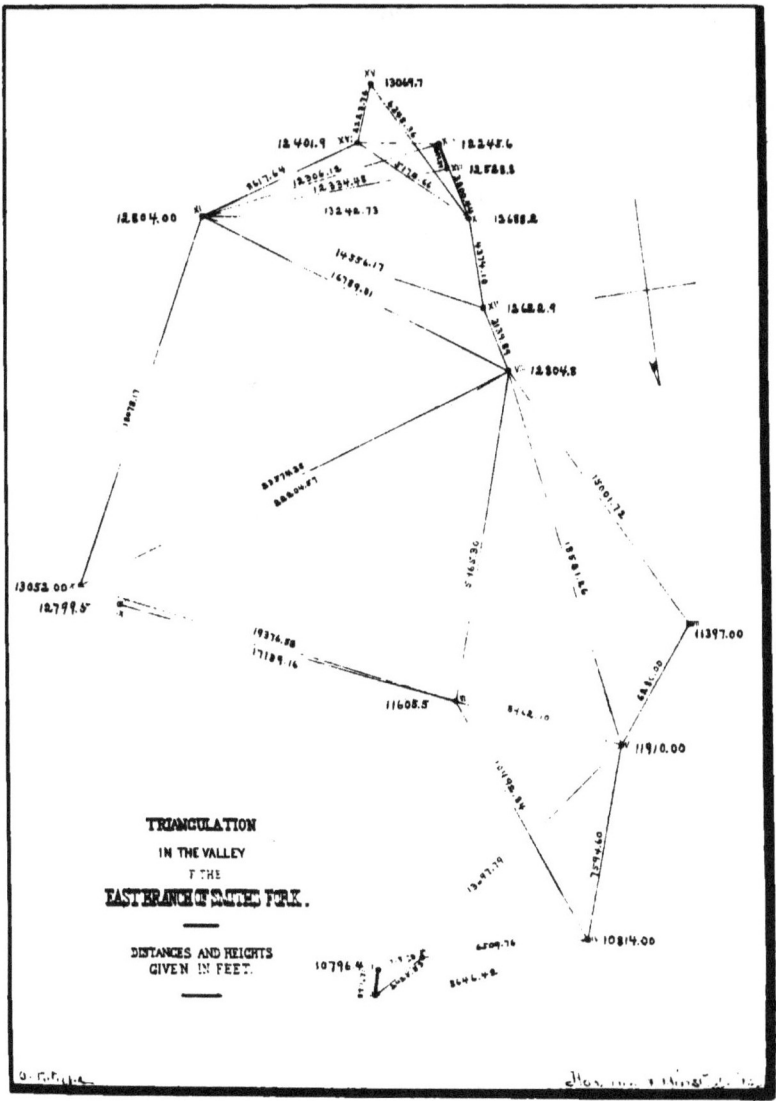

TRIANGULATION

IN THE VALLEY

F THE

EAST BRANCH OF SMITHS FORK.

DISTANCES AND HEIGHTS
GIVEN IN FEET.

to the very precipitous sides of the valley—and others not frequently seen elsewhere. These we have represented according to our best understanding of them, and it is hoped that the larger map, together with what is here written concerning the valley, will need no further explanations. A second map has been prepared, in which are embodied the results of our sketches on our road back to Fort Bridger, together with such information as we could obtain from the inhabitants, and an old map which Judge Carter was kind enough to lend us, which latter, however, we found incorrect in some particulars. This map we give as a mere hint as to the relation between Fort Bridger and the Smith's Fork Valley. The third map contains the skeleton triangulation of the main map. In this map the distances and heights are given in feet. As the heights have been given already we will only here repeat the distances :

Base-line, 897·71	(6–8) 15965·30	(11–16) 8617·64
(1–3) 1918·73	(7–8) 15001·72	(11–17) 12334·48
(2–3) 2425·53	(6–9) 17139·16	(8–12) 3139·89
(3–4) 6509·76	(6–10) 19376·58	(12–13) 4374·10
(2–4) 8646·42	(8–9) 22204·57	(13–14) 3800·44
(3–5) 13697·79	(8–10) 23574·38	(13–15) 6348·36
(4–5) 7594·60	(10–11) 18478·17	(13–16) 5170·66
(4–6) 10492·84	(11–8) 16789·81	(15–16) 2223·76
(5–6) 8462·10	(11–12) 14556·17	(14–17) 1299·89
(5–7) 6280·00	(11–13) 13242·73	(14–16) 3966·87
(5–8) 18581·26	(11–14) 12306·12	

A noticeable feature of this region is the scarcity of birds, reptiles, and insects. We saw only a few snakes and lizards on these plateaus. Musquitoes are very numerous in the months of June, July, and August, after which, we understand, they disappear. The deer-fly is a ferocious insect, very much like the common horse-fly, but larger and with irridescent colors, sometimes yellow, green, and red being banded together on its head. They go in swarms, attacking both man and beast, and are a terrible scourge to animals.

We have chosen to name one of these peaks (our point No. 11) Santa Anna Mountain, from our faithful Mexican guide.

HYPSOMETRICAL REPORT.

HYPSOMETRICAL REPORT.

I.—COLORADO.

OUR hypsometric work was entirely done by means of mercurial barometers. In Colorado this was done very hurriedly as to time, but not as to carefulness. We measured Pike's Peak, Mt. Lincoln, Mt. Bross, Mt. Silverheels, Mt. Quandary, Mt. Princeton, Mt. Gray, and Mt. Evans, with very fair results, as will be seen by the comparison which we make with the heights of these same peaks as given by other observers. We give this comparison in a tabular form, with the height of the two bases measured by us, as well as their heights as given by others:

	Princeton.	Hayden.	Wr. and Wy.*
Manitou	6351·00	6357·00	....
Pike's Peak...............	14147·28	14146·56†	....
Alma......	10364·50	....	Wr....10254·00
Mt. Bross...............	14255·20	....	....
Mt. Lincoln...............	14297·80	14297·00	Wy....14307·00
Mt. Silverheels............	13861·00	13897·00	Wy....13832·00
Mt. Quandary.............	14281·00	14269·00	....
Mt. Princeton........... ...	14208·90	14196·00	Wr....14041·00
Mt. Evans...............	14353·30	14340·00	....
Mt. Gray................	14363·30	14341·00	Wy....14319·00

* Wr., Wheeler. Wy., Whitney.   † U. S. Signal Service.

In many cases our measurements vary no more than ten feet either way from the others, and this small difference is to be considered as of no importance in such a height, as it would seem like disputing about the thousandth of an inch in a man's measurement.

Where no base is mentioned in the above table, we used the place nearest the mountain that we wished to measure ; obtaining its height by corresponding observation between it and some other place whose height was already known. For example, we obtained the height of our barometer in

Manitou by observations between it and the depot of the Rio Grande Railroad at Colorado Springs. The station in Manitou was in the Beebee House, in a room on the second floor on the rear of the building. In the height given, this difference is allowed for, and the elevation is that of the ground. The difference between our height and that of Dr. Hayden's may be due to the difference in the location chosen as the point to be measured, as this could easily be allowed for on such rough ground. In measuring Mts. Lincoln, Bross, Silverheels, and Quandary, we used Alma as a base, and obtained its height by a careful series of observations between it and a point the height of which was known, in Fairplay. For the measurement of Mt. Princeton, we used the height of Helena, as given by the surveys of the South Park Railroad. For the measurements of Mts. Gray and Evans, we took the height of the Clear Creek Railroad station at Georgetown as a basis, and measured the height up to our stationary barometer, which was in the Barton House. In the measurement of these last two mountains we compared the two mercurials and one aneroid very carefully, and left the aneroid below with an observer, Mr. Osborn, and took the others to the summits of the two mountains at the same time. It was our object to make simultaneous observations, in order to find out which of the two was the higher. In this we were successful, but were rather disappointed in the day being stormy, and thus not giving us as good an opportunity as we wished.

Our barometers were our constant companions, although we rode several hundred miles on horseback, and we usually carried in addition to them the rest of our instruments strapped to our bodies. We usually carried, besides the barometer, a pocket-sextant, aneroid, level, thermometer, clinometer and compass, and a telescope, together with our notebooks.

II.—UTAH.

Most of our barometric stations in our Utah measurements were also the stations of our triangles, and have therefore been already described in the topographical part of this

Western Slope of Continental divide,
Twin Lake Range.

report. Our barometers were always used in the same way as in Colorado, the same rules being followed in every case. We have reserved this matter, however, for the following section, where we intend to discuss the methods we made use of. It will be needless to repeat here the list of heights already given in the topographical report. The only peculiarity of our work in Utah was that the lower barometer was situated for most of the time in the centre of the valley of the East Branch of Smith's Fork.

Our usual method of procedure was, on reaching the top of the mountains, to lay the instruments on the ground and build a monument, which not only served as a point of reference for our transit, but as a support for the barometer. Then setting up the transit, and hanging the barometer by the side of the monument, we first made the transit observations, and then those of the barometer; thus, giving the barometer ample time to cool off and adjust itself, observations were then made on it, which lasted from half to three quarters of an hour. The mean of these was taken in the general calculations, but when greater care seemed to be needed, each observation was calculated separately.

We resorted to every expedient within our reach to make our work correct. We remeasured the peaks of whose measurements we were in any degree doubtful. When ascending one slope, if by means of our pocket-level we found we were on a level with a summit on the other, we measured the point on which we stood, and drew a sketch from it, which gave us a good contour line to follow.

Sketches were also made from the various peaks, which gave us the drainage of the valley, and of its neighboring ones, Black's and Henry's Forks. Our transit measurements were generally checked by the pocket-sextant, in what is known as a "tour d'horizon." We also made many measurements of the dip of the strata, where any importance was attached to them.

In addition to the instruments carried in Colorado, we had our transit with its tripod, and this made quite a load.

3

III.—BAROMETRIC WORK.

The barometers used were always carefully compared before and after each day's work, and a long comparison, lasting for three days, was made before and after our trip, which gave very good results, and proved the barometers worthy of confidence. There was a constant difference of ·002 of an inch, which was obtained from the observations made before we started. This was, at times, increased to ·004 of an inch, but never more.

We observed the usual elements for barometric measurements in the following order: date, time, temperature of the barometer, height of the barometric column, temperature of the air, and the general state of the atmosphere. The meteorological record will give the elements used in our corresponding observations. We also made it a point to expose our barometers to the free action of the atmosphere, but out of the direct rays of the sun, in the shade either of a monument, a house, or a tree. In all our measurements the lower barometer was brought as close as possible to the base of the mountain, in order to avoid any differences of pressure which might arise from the horizontal distance of the two instruments. For the same reason we preferred, when practicable, a position in a north and south line, in order to diminish the influence of barometric waves passing from west to east. The exact altitude of this lower station was always obtained (when we could not get the height from any reliable source) by a series of observations between it and some place the elevation of which could be relied upon. We were often helped in this by the consultation of the maps and surveys of railroads, to which we were given access in the most obliging way. The tables used in the calculation of the heights from the observations were the Hypsometrical Tables in the "Smithsonian Meteorological and Physical Tables," prepared by Dr. Guyot.

The influence of the time of the day at which the observations have been taken, an element too often neglected in barometrical measurements, has been corrected by means of a manuscript table by the same author. This table gives the corrections to be applied to the results obtained for each hour of the day, and each month of the year. These correc-

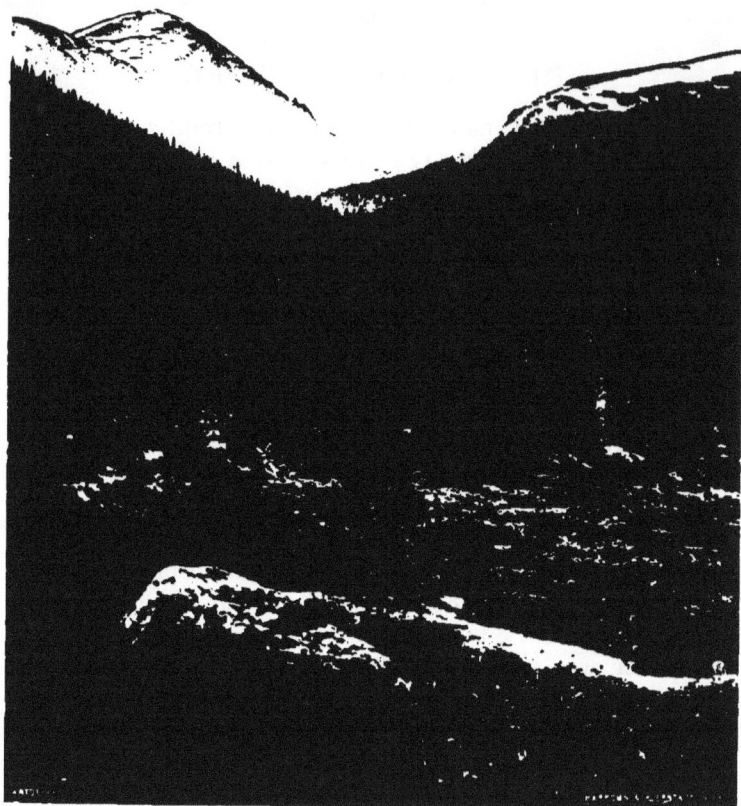

Blue River and Mountains, from
Hoosier Pass.

tions have been derived from the elaborate discussion by Prof. E. Plantamour of the long series of thermometrical and barometrical observations made at Geneva and the St. Bernard, as a means of ascertaining altitudes, in the "Mémoires de la Société de Physique de Genève." They are calculated for use in connection with the class of hypsometric tables founded upon the barometric formula of La Place, to which Dr. Guyot's tables belong.

The total correction thus applied for the influence of the hour of the day, contains two main corrections, one for the difference in amount and in time, above and below, in the daily tide of the barometer, the so-called *horary* (hourly) variation ; and another, by far the largest, for the error in the true mean temperature and moisture of the layer of air between the upper and lower barometer, as given by the observation of the instruments at both stations. As the last depends upon conditions of the atmosphere extremely variable, which cannot be reduced to a formula, it is evident that the amount of the total correction has to be modified according to the various states of the atmosphere existing at the time of the observation. When the sky is clouded, or during a fog, and with a gentle wind mixing the layers of the atmosphere, the average temperature of the air, given by the arithmetical mean of the upper and lower thermometers, will be much nearer the true one than when the air is calm, the sky clear, and the insolation and radiation excessive. In the latter case the layers nearer the ground, in which the instruments are necessarily placed, show abnormal temperatures, which, according to the hour of the day at which they are taken, are either above or below the mean temperature of the whole mass of the air.

We have, therefore, somewhat modified the amount of the correction when it seemed advisable to do so, after a careful consideration of the atmospheric circumstances indicated in the accompanying meteorological record.

We feel confident that by this method, the greater part, if not the whole, of the influence of the hour of the day is corrected.

The elements of the corrections having been derived from observations made between 1000 and 9000 feet, it remains to be seen whether they are equally applicable to higher altitudes. This matter remains for future investigation. Thus .

far the results of our measurements up to nearly 14,400 feet
seem to justify their use to an altitude of 15,000 feet.

We will now devote a little space to the aneroid barometers, and a few facts that were noticed concerning them during our stay in Colorado. The best one of these instruments
was given to the Palæontologists when we reached Fort
Bridger, and the other was left in our camp while we were
in the mountains, and used but little.

They were taken more for the purpose of giving them a
series of severe tests than for any work that we expected of
them. One or the other of them was always carried to the
top of each mountain we measured in Colorado, having been
previously carefully set and compared with the mercurial
barometers. In the measurement of the eight mountains upon
which they were used they always acted in the same peculiar
manner, seldom reaching their true height within half an
hour of the time they were on the top ; and on the return
having as much hesitation about coming to a standstill.

There seem to be two great causes of such actions. The
first and principal one is the unequal expansion and contrac-
tion of the metallic box upon which the measurements
depend ; the second of these causes is the mechanical work
performed by the same box in the rotation of the index.

This last difficulty can be, and in fact has been, overcome
in that form of aneroid made by Goldschmid, of Zurich;
but the first cannot be as easily cured, as it seems to be a
defect in the system employed. The metal having once been
expanded can never return to the same condition, a change
having probably taken place in the situation of its molecular
components.

We give the following as examples of this action of the
aneroids ; they are a fair sample of the rest, and not exagger-
ated ones :

Pike's Peak, July 5th, 1877.—Barometer 2233, at Lake House, two feet above the ground.

TIME.	T.	B.	t.	Aneroid.	REMARKS.
A.M.					
5.40	58.2	21.014	59.0	21.38	*Fair and warm.*
5.50	60.5	21.014	59.9	21.39	*Clouds 1. cirrus.*
5.60	61.6	21.016	60.0	21.39	*Wind N. W. 1.*

Pike's Peak, July 5th, 1877.—Baromete, 2233, on the outside of the house of the Signal Service. Two feet six inches below the highest point of the mountain.

Time.	T.	B.	t.	Aneroid.	Remarks.
A.M.					
9.45	53.6	18.250	49.1	18.85	*Fair and warm.*
9.50	52.8	18.252	48.4	18.83	*Clouds 4, cirro cumulus.*
9.55	54.3	18.250	49.1	18.85	*Wind S. E. 1.*
10.00	52.3	18.250	47.3	18.85	
10.05	50.3	18.248	48.1	18.85	
10.10	51.6	18.248	48.3	18.89	
10.15	50.6	18.250	47.1	18.83	

This table also gives the form in which our observations were taken and kept.

These facts lead us to infer that, though the aneroid may answer very well for the comparatively small heights of our Eastern coasts, it will hardly do for such great elevations, or for such long distances as are found in the West. It needs constant watching and correcting by comparisons with a mercurial barometer, as its zero point is apt to change by the handling of the instrument. The influence of the temperature, even in those which bear on their face " *compensated*," is by no means eliminated, and a correction has to be found experimentally for each instrument. It also needs (an idea which few persons having an aneroid seem to realize) considerable care in transportation, and an accident to it is wholly beyond the possibility of repair in the field, while a mercurial barometer, with the aid of a little ingenuity, can be readily repaired, if pains have been taken to carry the requisite duplicate parts along with the instrument.

METEOROLOGICAL REPORT.

UTAH.

Mt. Lincoln and the Head Waters

of the Platte.

METEOROLOGICAL REPORT.

THE six weeks which were spent in the mountain valley of Smith's Fork were marked by great variations of temperature, and many curious phenomena, both in atmospheric and magnetic changes, which were very noteworthy. We were somewhat sheltered in our camp from the violence of the storms which swept over the mountains above us, but at the last part of our stay we had the only alternative of working early in the morning and on till eleven o'clock A.M., or doing no work at all. When it is also remembered that a height of 3000 feet had to be climbed from our camp before our real day's work began, and that, even then, we were not at all sure that we would not be driven out by electric storms at any moment, some of the difficulties of our work may be appreciated. The storms, usually of snow or hail, set in on the mountains at eleven and continued until late in the afternoon, during which time it was almost an impossibility to go near the top of the ridge without the most painful sensations caused by electricity. While the hailstorms lasted (and they were frequent) those who were on the ridge had to seek shelter under the ledges to avoid the hard blows from globes of ice often reaching from one half to three quarters of an inch in diameter. We were finally driven from the mountains by these storms, as they reduced our working hours to a minimum ; and we had to return to Fort Bridger. The meteorological record, though a mere record of the state of the atmosphere for a few days at a time, gives, nevertheless, on a close examination, some very interesting facts in regard to the curious action of the air; but it can be of no great value otherwise. The observations were taken every fifteen minutes during the days in which they were kept, and were made with more special reference to the corresponding observations of the barometer, in order to obtain the full account of the circumstances under which we were working, than for a regular and systematic meteorological record. The tables explain themselves, giving in separate columns the various necessary elements, viz.: Date, time, temperature, barometer observed, barometer reduced to the freezing point, wet and dry bulb thermometers, their difference and the relative humidity derived from it, the wind's direction and force, the cloudiness and the nature of the clouds, with some remarks on the state of the weather.

The storms which we encountered were without doubt the result of the overheating of those large and elevated plateaus, the inclined sides of which, as well as the mountain slopes, expose a larger surface to the rays of the sun than would a merely flat body. These circumstances would seem to help the formation of strong ascending currents, which, having once been started, furnish ample material for the midday showers. These storms gradually extend their limits, as to time, until the rainy season is complete. They seemed to be similar to those of tropical regions, and this fact enabled us to explain the presence of such large quantities of electricity in the air. Once only, during our stay, did we observe two different layers of clouds, one above the other, moving in opposite directions. Their line of motion was almost due N. E. and S. W., the upper one moving S. W. This was seen from the cliff on which No. 9 is situated, and was watched as a great curiosity, as one cloud was almost on a level with us, and the other some distance above. We were now and then favored with beautiful rainbows, on two occasions seeing triple arcs, and a single one was the exception rather than the rule. The temperature during the first part of our stay was quite favorable to our work, being just cool enough for the exercise of brisk walking during the day; but at night generally freezing the edges of the brook near our camp, or any water left out in small vessels, thus making a number of blankets very necessary articles for our comfort; and a blazing camp-fire of pine stumps not only a pleasure to the eyes, but to the fingers also. Before the end of our stay (our camp was at a height of 10,500 feet), we needed all our heavy clothing and generally a fire all day as well as at night.

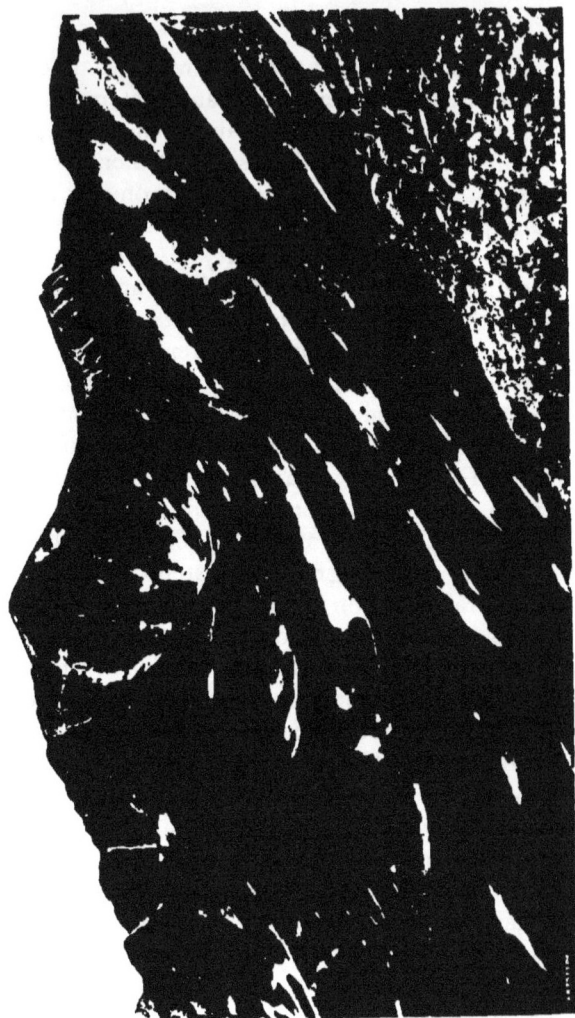

The electricity spoken of above was very troublesome, rendering the needle of the transit almost useless, and running in a stream at times from the end of the barometer to the ground. We received a great number of shocks when using the instruments. The discharge seemed to come from above us, but was not always accompanied by strokes of lightning. We found that the simple practice of lifting the hand above the head with one finger elevated served to relieve the unpleasant sensation of buzzing behind the ears, and, when once more it became unbearable, by lifting another finger we made the body a sort of doubled Leyden jar. The delay thus obtained enabled us to pack the instruments with the other hand and to get away from the monument. In one instance the monument (ill-fated No. 18, which was only just built) was torn into fragments by lightning while we were but a short distance off.* We sometimes found the wind very troublesome while making observations with the transit, as it blew with terrific force over the exposed summits of the peaks. We were, however, very much pleased with the invigorating air of the place, and when it was clear we could not have wished for greater enjoyment in out-door life than we found there.

* U. S. Geol. and Geog. Survey, 1874, p. 456, etc.

Barometer 2219, at Station in Camp Nassau. (3.)

DATE.	A.M.	T.	B.	32° B.	D. B.	W. B.	D −	R. H.	Wind, direction and force (1 – 10).	Clouds (1 – 10).	Nature of clouds.	REMARKS.
Thursday, August 9th, 1877.	8.30	68.4	29·654	+·582	64.1	43.7	20.4	38.1	N. E. 2	0	......	Fair; warm.
	9.	70.1	29·656	+·580	66.2	46.1	20.1	40.2	E. 2	0	......	"
	9.30	71.8	29·656	+·576	67.6	45.6	22.1	41.6	S. E. 2	0	......	"
	10.	70.5	29·656	+·580	65.9	44.7	21.2	42.0	E. 2	0	......	"
	10.15	70.3	29·657	+·581	67.4	46.3	21.1	45.2	E. 1	0	......	"
	10.30	70.9	29·656	+·580	66.7	46.7	20.0	40.2	E. 1	0	......	"
	10.45	70.6	29·658	+·582	68.8	47.5	21.3	46.8	E. 1	1	......	"
	11.	71.3	29·658	+·580	68.4	47.5	20.9	39.9	N. E. 3	1	......	"
	11.15	69.5	29·657	+·583	66.3	45.8	20.5	39.7	N. E. 2	2	Cirrus.	"
	11.30	69.7	29·656	+·582	66.4	46.4	20.0	40.2	E. 2	2	"	"
	11.45	69.8	29·655	+·581	66.5	47.5	19.0	42.7	E. 1	3	Cirro-cumulus.	Very warm.
	12.	69.6	29·651	+·577	67.9	47.9	20.8	40.7	E. 1	4	"	"
	P.M. 12.15	71.7	29·651	+·573	68.5	47.7	20.2	40.7	E. 1	4	"	Warm.
	12.30	72.3	29·651	+·571	66.7	47.3	19.4	42.7	S. E. 2	5	"	"
	12.45	72.3	29·649	+·569	67.8	48.2	19.6	43.2	S. E. 2	6	"	"
	1.	74.1	29·649	+·566	67.8	47.6	20.2	40.7	S. E. 3	6	"	"
	1.15	73.2	29·650	+·568	68.8	48.6	20.3	41.3	E. 3	6	Cumulus.	"
	1.30	76.6	29·654	+·563	70.5	50.2	16.7	42.3	N. E. 3	7	"	"
	1.45	77.4	29·648	+·567	69.3	53.6	17.6	27.0	N. E. 3	7	"	"
	2.	72.1	29·640	+·570	69.3	51.7	11.9	21.8	E. 2	8	"	but pleasant.
	2.15	70.1	29·635	+·564	61.4	49.5	15.6	37.2	N. E. 4	8	"	"
	2.30	69.7	29·633	+·564	65.3	49.7	16.6	23.6	N. E. 4	8	"	"
	2.45	67.7	29·632	+·559	66.3	49.7	18.4	20.7	N. E. 4	8	"	"
	3.	67.8	29·628	+·561	66.2	48.8	16.7	13.8	E. 2	6	"	"
	3.15	67.9	29·625	+·557	65.5	47.8	14.8	19.3	S. E. 2	7	Cirro-cumulus.	"
	3.30	66.3	29·623	+·556	65.2	47.6	15.1	24.0	S. E. 3	7	"	"
	3.45	67.3	29·622	+·553	64.6	50.1	14.0	26.4	S. E. 3	7	"	Cooler.
	4.	66.7	29·619	+·552	64.2	47.6	15.2	25.7	S. E. 3	7	"	"
	4.15	66.1	29·620	+·550	63.9	49.0	14.9	25.1	S. E. 3	7	"	Thunder in S. E.
	4.30	65.9	29·617	+·553	62.8	49.0	15.3	26.7	S. E. 3	7	"	Cool.
	4.45	64.6	29·617	+·552	63.8	47.5	16.2	22.4	S. E. 1	6	Cumulus.	"
	5.	65.8	29·616	+·550	63.3	47.6	15.5	19.3	S. W. 1	7	"	"
	5.30	65.2	29·616	+·549	62.5	47.8	15.5	20.8	S. 1	6	"	"
	6.	63.0	29·610	+·554	63.3	47.0	13.4	20.8	S. 1	6	Cirro-cumulus.	"
	6.30	61.7	29·612	+·552	59.9	46.5	—	28.1	S. E. 1	6	"	"
	7.	55.3		+·563	54.9	45.2	9.7	41.1		5	"	"

T. = Attached thermometer. B. = Barometer observed. 32° B = Reduced to freezing-point. D. B. = Dry bulb thermometer. W. B. = Wet bulb thermometer.
D. − = Difference between dry and wet thermometer. R. H. = Relative humidity.

Barometer 2219, at Station in Camp Nassau. (3.)

DATE	A.M.	T.	B.	32° B.	D. B.	W. B.	D —	R. H.	Wind, direction and force (1 – 10).	Clouds (1 – 10).	Nature of clouds.	REMARKS.
Friday, August 10th, 1877.	9.	67·8	20·689	+·618	63·4	48·2	15·2	23·8	S. 1	2	Cirrus.	Fair now, but heavy clouds [earlier.
	9.30	68·6	20·691	+·619	65·6	50·2	15·4	26·4	S.E. 2	4	Cirro-cumulus,	Cooler.
	10.	67·8	20·692	+·621	62·4	48·7	13·7	28·8	S.E. 2	6	" "	Cool.
	10.15	65·8	20·692	+·625	61·0	48·1	13·4	29·3	S. 2	8	" "	" "
	10.30	63·5	20·684	+·621	60·6	48·8	12·4	34·2	S. 3	8	Cumulus.	" and breezy.
	10.45	68·9	20·692	+·620	64·6	48·3	15·8	23·6	S. 3	8	"	" "
	11.	68·0	20·694	+·623	60·2	49·3	15·6	29·3	S. 3	8	"	" "
	11.15	65·2	20·690	+·625	57·8	47·1	13·1	29·3	S. 2	5	"	" (Rain).
	11.30	62·1	20·683	+·619	60·5	46·4	11·4	29·3	S. 2	7	"	Warm.
	11.45	68·8	20·680	+·616	65·1	47·5	13·0	34·0	S. 3	7	Cirro-cumulus.	" "
	12.	59·9	20·688	+·616	63·4	51·7	13·4	22·4		8	" "	" "
	P.M.											
	12.15	67·9	20·690	+·618	63·6	47·9	15·5	31·7	S.E. 3	8	" "	" "
	12.30	72·6	20·689	+·614	67·8	49·9	17·3	20·6	S. 3	8	" "	" "
	12.45	69·4	20·694	+·620	64·4	50·5	15·2	25·1	S. 3	8	" "	" "
	1.	70·1	20·694	+·618	64·1	49·2	13·8	32·9	S. 2	7	Cumulus.	Cooler.
	1.15	67·5	20·688	+·617	62·4	50·3	13·7	30·6	S.W. 3	8	Cirro-cumulus.	" "
	1.30	64·6	20·684	+·619	61·8	48·7	13·7	30·6	S.W. 3	8	" "	" "
	1.45	61·9	20·683	+·623	58·8	48·4	13·1	36·1	W. 3	9	" "	Breezy.
	2.	60·7	20·681	+·621	58·2	46·9	11·9	33·8	W. 4	9	Cumulus.	Quite cool.
	2.15	60·4	20·681	+·623	58·3	46·3	11·9	35·8	W. 4	9	" "	" "
	2.30	58·4	20·681	+·627	56·1	46·9	11·4	36·7	W. 3	9	" "	" "
	2.45	58·5	20·682	+·628	56·4	45·3	10·8	36·7	S.W. 4	9	" "	" "
	3.	58·2	20·682	+·628	55·7	45·1	10·6	38·8	S.W. 4	4	Cumulus nimbus.	" "
	3.15	63·4	20·680	+·616	61·7	45·8	12·9	35·3	S.W. 3	3	Cirro-cumulus.	Pleasant.
	3.30	65·7	20·679	+·613	61·6	49·6	12·0	38·1	S. 2	6	" "	" "
	3.45	62·2	20·678	+·617	60·3	48·9	11·4	39·2	S. 2	7	" "	" "
	4.	61·7	20·674	+·613	59·8	49·2	11·1	38·1	S.W. 3	8	" "	" "
	4.15	60·6	20·670	+·610	58·7	48·7	11·6	37·0	S. 2	3	" "	Cool.
	4.30	60·4	20·667	+·609	57·8	47·1	11·0	37·0	S. 2	5	" "	Pleasant.
	4.45	58·8	20·665	+·607	57·8	47·6	11·6	38·8	S. 1	2	" "	" "
	5.	56·8	20·664	+·610	50·0	46·2	11·6	39·8	S. 1	2	" "	Cool.
	5.30	54·4	20·664	+·614	54·2	45·9	10·1	45·8	E. 2	1	" "	" "
	6.	53·3	20·664	+·617	53·4	45·4	8·8	48·3		3	Cirrus.	
	6.30			+·619		45·3	8·1		 1			
	7.											

Barometer 2219, at Station in Camp Nassau. (3.)

DATE.	A.M.	T.	B.	32° B.	D. B.	W. B.	D —	R. H.	Wind, direction and force (1–10).	Clouds (1–10).	Nature of clouds.	REMARKS.
Saturday, August 11th, 1877.	9.	61.0	20.724	+.663	57.0	49.0	8.0	52.1	E. 2	8	Cumulus.	Cool.
	9.15	61.0	20.718	+.657	59.0	49.0	10.0	43.2	E. 2	8	"	Few rain-drops.
	9.30	58.8	20.718	+.658	56.0	48.5	8.5	53.6	N.E. 2	9	"	" "
	9.45	58.0	20.716	+.662	55.2	47.0	8.2	50.2	N.W. 3	9	"	Light shower.
	10.	55.0	20.716	+.667	53.6	45.5	8.1	48.3	N.W. 4	9	"	Cool and threatening shower.
	10.15	55.0	20.716	+.667	53.2	46.0	7.2	53.1	N.W. 3	9	"	" " thunder in the dis-
	10.30	55.5	20.718	+.668	53.0	45.8	7.2	53.5	Calm.	9	"	" " { tance.
	10.45	55.0	20.716	+.667	53.8	46.0	7.8	51.8		9	"	
	11.	55.0	20.716	+.667	53.0	47.0	6.0	60.6		9	"	showers S. W.
	11.15	54.5	20.714	+.666	53.2	45.2	7.8	50.8	"	9	"	Clearing in W.
	11.30	55.5	20.714	+.664	53.2	47.0	6.2	60.6	"	8	"	"
	11.45	57.0	20.714	+.663	55.0	47.5	7.5	52.7	W. 2	8	"	Rain E.
	12. P.M.	56.0	20.712	+.663	54.5	45.0	9.5	41.1		8	"	"
	12.15	55.0	20.710	+.662	54.0	45.0	9.0	43.4	W. 2	8	"	"
	12.30	54.5	20.708	+.661	52.5	45.2	8.8	45.8	W. by S. 3	8	"	Cloudy in W.
	12.45	54.0	20.706	+.658	53.5	45.0	7.5	50.8	W. 2	8	"	"
	1.	54.5	20.706	+.658	53.0	45.2	8.0	51.8	W. 2	9	Stratus.	"
	1.15	54.5	20.706	+.660	52.0	46.0	6.8	48.3	Calm.	9	"	and cool
	1.30	53.5	20.704	+.659	54.0	45.2	8.8	56.2		10	"	"
	1.45	53.0	20.700	+.657	52.0	45.2	8.0	49.3		8	"	Rain.
	2.	53.0	20.700	+.653	51.8	46.0	7.8	49.8	"	8	"	Clearing in W
	2.15	53.4	20.698	+.647	54.5	44.2	8.0	46.1	W. 3	8	Cumulus and str.	"
	2.30	56.6	20.694	+.644	54.6	43.8	9.5	41.1	S.W. 3	6	" "	"
	2.45	55.8	20.690	+.636	55.0	45.0	9.8	39.9	S.W. 3	6	" "	"
	3.	57.5	20.688	+.636	55.2	44.8	8.6	46.9	S.W. 2	6	" "	"
	3.15	56.6	20.686	+.632	54.8	46.4	9.0	43.4	S.W. 4	8	" "	Cool.
	3.30	57.8	20.682	+.632	54.5	45.2	8.6	46.9	S.W. 2	9	Cumulus	Cloudy and cool.
	3.45	57.2	20.682	+.637	50.0	46.2	8.9	45.8	S.W. 4	9	"	Threatening and cool.
	4.	52.8	20.682	+.637	50.0	45.0	5.0	64.6	S.W. 3	9	"	"
	4.15	52.5	20.684	+.643	52.6	45.2	4.8	67.6	S.W. 2	9	and str.	Rain, light.
	4.30	50.8	20.688	+.649	49.2	45.0	5.2	63.8	S.W. 2	10	" "	" "
	4.45	50.0	20.688	+.647	49.5	45.2	4.5	67.0	S.W. 2	10	" "	" "
	5.	50.8			49.6	45.0	4.6	67.0	N. 3	10	" "	" "
	5.30				47.0	44.2	2.8	80.7	S.W. 2	8	" "	
	6.				44.5	44.0	.5	63.8			" "	

Barometer 2219, at Station in Camp Nassau. (4.)

DATE.	A.M.	T.	B.	32° B.	D. B.	W. B.	D —	R. H.	Wind, direction and force (1 - 10).	Clouds (1 - 10).	Nature of clouds.	REMARKS.
Tuesday, August 14th, 1877.	9.	57·8	20·700	+·648	56·4	46·9	9·5	42·2	N.W. 2	3	Cumulus.	Fair, warm.
	9.30	59·4	20·700	+·644	56·3	46·5	9·8	42·5	N.W. 3	5	"	" pleasant breeze.
	10.	56·8	20·700	+·650	53·8	45·3	8·5	45·8	N.W. 2	8	"	Clouded over, cool.
	10.15	55·8	20·697	+·648	54·4	45·6	8·8	45·8	W. 3	9	"	" " "
	10.30	59·8	20·701	+·645	56·7	46·2	10·5	37·9	W. 3	9	"	" " "
	10.45	57·7	20·700	+·646	55·2	45·1	10·1	38·8	W. 4	8	"	" " "
	11.	58·4	20·700	+·646	55·4	45·6	9·8	41·1	W. 5		nimbus.	" " "
	11.15	59·6	20·700	+·644	55·7	46·1	9·6	42·2	W. 3	7	"	Rain, clouded over, cool.
	11.30	59·5	20·698	+·642	56·8	46·5	10·5	39·9	S.W. 3	6	"	Hail storm, 10 minutes.
	11.45	58·4	20·697	+·643	58·8	48·3	10·5	40·1	S.W. 3	4	"	Pleasant.
	12.	60·5	20·700	+·642	60·4	47·7	12·7	31·2	S.W. 2		"	"
P.M.	12.15	61·3	20·700	+·640	59·5	46·5	13·0	28·1	W. 3	4	"	"
	12.30	63·3	20·700	+·633	59·8	47·7	12·1	33·0	N. 3	6	"	"
	12.45	63·4	20·701	+·634	60·2	47·3	12·9	31·9	N. 2	6	"	"
	1.	58·2	29·699	+·645	56·4	45·3	11·1	34·6	N. 5	5	stratus.	"
	1.15	59·6	20·700	+·644	57·7	46·2	11·5	33·8	W. 3	2	"	"
	1.30	60·8	20·697	+·642	59·1	46·7	11·4	35·8	N.W. 2	1	nimbus.	"
	1.45	61·1	20·698	+·637	59·4	46·4	13·0	28·1	N. 3	3	"	Clouding over, fast.
	2.	62·2	20·697	+·637	60·8	48·3	12·5	32·4	W. 3	7	"	Cool.
	2.15	61·6	20·692	+·637	57·8	45·2	12·6	28·6	W. 3	9	"	"
	2.30	57·7	20·693	+·641	56·2	44·6	12·0	39·2	W. 4	9	"	"
	2.45	57·8	20·694	+·631	56·4	44·4	12·0	31·2	N.W. 3	3	"	"
	3.	58·8	20·691	+·637	56·6	44·9	11·1	31·2	N.W. 3	5	"	"
	3.15	58·8	20·689	+·637	58·4	45·3	13·1	26·7	N.W. 3	5	"	"
	3.30	57·0	20·685	+·630	55·3	44·1	11·2	33·3	N.W. 3	3	"	"
	3.45	59·4	20·682	+·634	58·3	45·8	12·5	28·6	N.W. 3	5	"	Warm, sun out.
	4.	59·3	20·680	+·634	59·7	43·4	14·3	22·9	W. 3	5	"	Cool again.
	4.15	59·1	20·680	+·635	57·6	43·4	13·6	22·0	W. 3	4	"	"
	4.30	58·8	20·679	+·630	57·0	43·1	14·4	20·1	W. 3	5	"	"
	4.45	56·8	20·679	+·629	55·3	43·2	13·9	22·0	W. 3	7	"	Pleasant.
	5.	56·8	20·679	+·629	55·4	42·4	12·7	24·3	W. 4	6	"	"
	5.30	55·2	20·678	+·630	54·6	42·3	12·3	26·4	W. 1	2	stratus.	"
	6.	54·1	20·677	+·638	57·7	42·3	10·4	35·0	W. 1	2	Stratus.	
	6.30	49·8	20·676		49·2	41·4	7·8	46·6	S. 1	4	"	

Barometer 2219, at Station in Camp Nassau. (4.)

Date	A.M.	T.	B.	32° B.	D. B.	W. B.	D —	R. H.	Wind, direction and force (1 — 10).	Clouds (1 — 10).	Nature of clouds.	Remarks.
Wednesday, August 15th, 1877.	9.45	60·0	20·686	+·628	58·8	45·1	13·7	24·9	N. by W. 2	3	Cumulus.	Pleasant.
	10.	63·2	20·688	+·623	60·5	46·6	13·6	26·3	N. by W. 3	4	"	"
	10.15	62·8	20·688	+·624	58·8	45·2	13·6	24·9	N. by W. 3	5	" and nimbus.	Cloudy.
	10.30	59·2	20·690	+·632	57·0	44·1	12·9	27·2	W. 2	6	"	Little rain.
	10.45	57·0	20·690	+·636	55·8	44·5	11·3	33·3	S. 2	7	"	Clear.
	11.	62·8	20·692	+·628	60·5	47·6	12·9	31·2	W. 2	5	"	"
	11.15	61·8	20·694	+·633	58·8	45·6	13·2	26·7	W. 3	5	"	" and rain in E.
	11.30	60·2	20·694	+·632	60·0	47·0	13·0	29·3	N. by W. 3	5	"	" in E.
	11.45	60·5	20·694	+·636	60·4	47·0	13·4	29·3	N. by W. 3	6	"	" in E.
	12.		20·694	+·636	58·0	45·5	12·5	28·6			"	
	P.M.											
	12.15	63·8	20·696	+·630	61·5	47·0	14·5	24·0	W. 5	5	"	Showers in E.
	12.30	63·0	20·692	+·632	60·2	45·0	15·2	19·6	N. 4	5	"	" "
	12.45	58·8	20·690	+·636	55·5	43·4	12·1	27·8	W. 3	7	"	Little rain.
	1.	62·8	20·690	+·627	61·4	46·5	14·9	22·7	W. 4	6	"	Clear.
	1.15	65·0	20·688	+·623	63·8	44·6	20·3	2·6	W. 3	6	"	"
	1.30	66·2	20·688	+·619	60·5	43·5	18·9	7·9	W. 4	5	"	"
	1.45	61·8	20·686	+·627	58·8	45·2	15·3	19·6	W. 4	6	"	"
	2.	60·8	20·684	+·628	59·8	44·5	14·3	21·5	W. 3	6	"	Cloudy.
	2.15	60·5	20·682	+·626	61·5	45·8	14·0	22·9	W. 4	5	"	Clear.
	2.30	63·0	20·680	+·621	59·0	45·6	15·9	18·0	W. 5	4	"	"
	2.45	61·5	20·678	+·621	60·5	46·2	12·8	30·0	W. 6	4	"	"
	3.	62·0	20·678	+·618	61·0	46·0	14·5	22·7	W. 3	4	"	Cloudy.
	3.15	61·8	20·678	+·616	60·0	45·5	16·2	15·0	W. 4	3	"	"
	3.30	61·5	20·674	+·616	59·6	45·0	15·1	19·6	W. 3	5	"	"
	3.45	59·0	20·672	+·616	57·8	43·8	14·6	21·2	W. 3	4	"	Clear.
	4.	60·5	20·670	+·618	59·0	43·2	13·8	23·5	W. 4	4	"	"
	4.15	58·2	20·670	+·613	57·8	46·8	15·2	16·6	W. 3	3	"	Cloudy and cool.
	4.30	60·5	20·668	+·616	61·6	46·6	14·6	18·3	W. 4	2	"	Clear.
	4.45	60·0	20·668	+·612	61·0	44·8	14·2	24·3	W. 3	2	"	"
	5.	58·5	20·668	+·607	59·3	44·8	16·6	15·0	W. 3	2	"	"
	5.30	53·8	20·662	+·610	57·0	44·5	14·5	19·8	W. 2	1	"	"
	6.30	48·6	20·654	+·614	50·5	42·5	12·5	27·2	W. 3		"	"
	7.			+·619	48·3	40·1	8·0	45·0	W. 1		Stratus.	
				+·618			8·2	42·6				

Barometer 2219, at Station in Camp Nassau. (4.)

Date	A.M.	T.	B.	32° B.	D. B.	W. B.	D —	R. H.	Wind, direction and force (1 — 10).	Clouds (1 — 10).	Nature of clouds.	Remarks.
Thursday, August 16th, 1877.	9.	59.8	20.694	+.638	56.5	46.5	10.0	39.9	W. 1	8	Cumulus stratus.	Cool; cloudy.
	9.30	61.3	20.696	+.636	59.2	48.2	11.0	38.1	W. 2	7	" nimbus.	" "
	10.	60.2	20.699	.641	57.6	46.3	11.3	35.8	W. 3	10	" "	" "
	10.15	59.5	20.699	.643	57.7	46.4	11.0	35.8	W. 3	6		" "
	10.30	62.3	20.699	.638	59.9	47.3	13.5	26.3	W. 1	8		" "
	10.45	62.9	20.697	.636	60.2	42.4	12.9	31.2	W. 1	8	Cumulus.	" .
	11.	63.6	20.698	.635	61.6	44.6	19.2	3.3	W. 3	9	" nimbus.	" ..
	11.15	58.4	20.697	.643	56.3	44.9	11.7	31.2	W. 2	10	Nimbus.	Slight rain.
	11.30	57.0	20.695	.643	55.4	46.8	10.5	35.4	W. 1	10	" "	Cool; cloudy.
	11.45	58.6	20.694	.640	56.9	46.6	10.1	39.9	W. 3	10		" "
	12.	60.5	20.694	+.636	57.0		10.4	39.9	W. 3	12		" "
	P.M.											
	12.15	56.0	20.693	+.643	53.9	45.8	8.1	48.3	W. 2	10	" "	" "
	12.30	57.5	20.693	.641	52.8	45.6	7.2	53.5	W. 1	10		" "
	12.45	59.0	20.692	.636	52.4	45.4	7.0	53.5	W. 1	10		" "
	1.	50.6	20.687	.648	49.3	44.0	5.3	63.8	W. 2	10		Rain.*
	1.15	47.3	20.683	.649	47.2	43.4	3.8	73.2	W. 3	10		" " and wind,
	1.30	48.6	20.683	.647	48.3	43.8	3.3	77.7	W. 3	10		" "
	1.45	49.0	20.683	.645	49.2	46.1	3.1	78.1	N. W. 7	10	" "	Rain.
	2.											
			No observations; wind, rain, and smoke drove me out.										
	2.15	58.0	20.683	+.628	53.0	46.2	6.8	57.0	N. 6	10	" "	Cold.
	2.30	59.6	20.683	.627	54.5	47.5	7.0	55.2	N. W. 2	10	" "	Cold; rain.†
	2.45	54.0	20.698	.631	51.4	45.1	6.3	99.0	N. 4	10	" "	" "
	3.	49.4	20.696	.658	48.6	43.7	4.8	66.3	N. W. 5	10	" "	" rain.
	3.15	50.5	50.695	.656	48.5	43.3	2.8	89.0	N. W. 5	10	" "	" "
	3.30	47.8	20.685	.651	46.2	44.5	4.5	66.3	N. W. 3	10	" "	" rain.
	3.45	47.9	20.684	.650	49.0	47.5	2.3	67.0	N. S. 2	10	" "	" "
	4.	51.9	20.683	.642	49.8	46.8	4.6	85.3	S. E. 2	7	Cumulus nimbus.	Clearing.
	4.15	53.2	20.680	.632	51.4	46.1	6.2	68.3	S. E. 2	5	" stratus.	" "
	4.30	54.6	20.679	.631	52.3	46.9	6.0	59.8	E. 1	4	" "	Cool.
	4.45	53.4	20.676	.634	51.5	45.5	5.9	59.0	E. 1	8	Cumulus.	" "
	5.30	49.2	20.672	.626	52.8	44.3	5.8	62.6	S. E. 1	7	" "	" "
	6.	49.2	20.664	.619	50.3	43.9	4.6	61.1	S. E. 1	7	" "	" "
	6.30	40.3	20.657	.622	48.5	42.9	5.8	66.3		7	" stratus.	" "
	7.	48.8	20.660		48.7			59.4		9		" "

* Took barometer in the tent, two feet below station.   † Barometer out again.

Barometer 2219, at Station in Camp Nassau. (4.)

DATE	A.M.	T.	B.	32° B.	D.B.	W.B.	D —	R.H.	Wind direction and force (1 — 10).	Clouds (1 — 10).	Nature of clouds.	REMARKS.
Friday, August 17th, 1877.	10.45	66.5	20.710	+.639	63.5	48.5	15.0	23.8	W. 3	5	Cumulus.	Clear and warm.
	11.	67.5	20.708	+.636	65.2	47.8	17.4	16.5	W. 4	6	" and stratus.	" "
	11.15	67.5	20.708	+.636	64.0	48.5	15.5	22.2	N. W. 5	6	" "	" "
	11.30	65.8	20.708	+.639	63.0	49.0	14.0	28.3	N. W. 4	6	" "	" "
	11.45	66.0	20.708	+.638	62.5	47.8	14.7	24.0	N. W. 5	6	" "	" "
	12.	61.5	20.706	+.645	59.6	47.5	12.1	33.0	N. W. 5	6	" "	" "
	P.M.											
	12.15	60.5	20.702	+.642	58.0	47.5	10.5	39.0	N. W. 6	7	" "	Cloudy in W.*
	12.30	59.2	20.698	+.642	57.8	47.0	10.8	39.0	N. W. 5	8	" "	Cloudy; rain E.
	12.45	58.5	20.696	+.640	57.0	48.2	8.8	48.8	N. W. 3	9	" and nimbus.	" "
	1.	57.0	20.693	+.639	55.0	46.5	8.5	46.9	S. E. 4	10	" "	Shower.
	1.15	48.8	20.693	+.654	45.9	45.5	.5	96.0	S. E. 2	10	" "	Rain and hail.
	1.30	47.5	20.692	+.656	45.3	43.8	1.88	87.8	S. E. 1	10	" "	" "
	1.45	47.5	20.692	+.656	40.5	42.4	2.7	80.3	S. E. 3	10	" "	Sprinkling.
	2.	47.5	20.692	+.656	47.0	44.5	4.6	65.7	S. E. 3	10	" "	" "
	2.15	48.0	20.692	+.656	44.5	45.5	4.0	100.0	S. E. 4	10	" "	Very cloudy.
	2.30	49.8	20.692	+.653	50.5	45.5	0.0	64.6	S. E. 3	10	" "	Sprinkling.
	2.45	50.2	20.692	+.651	50.0	47.0	3.0	78.6	S. E. 3	10	" "	Very cloudy.
	3.	50.5	20.694	+.653	48.5	45.2	3.3	77.7	S. E. 3	8	" "	" "
	3.15	53.2	20.690	+.645	52.0	47.2	4.8	68.9	S. E. 3	10	" "	Rain.
	3.30	53.2	20.688	+.643	52.0	47.2	5.2	68.9	S. E. 3	8	" "	Cloudy.
	3.45	53.4	20.680	+.635	52.3	47.0	5.1	66.0	S. E.	8	" "	Cool.
	4.	53.6	20.666	+.621	52.6	47.8	4.6	68.9	E. 2	7	" "	" "
	4.15	53.8	20.664	+.619	52.6	48.3	4.3	72.7	E. 1	7	" "	" "
	4.30	54.3	20.662	+.617	54.0	48.3	6.7	57.9	E. 1	6	" "	Cool.†
	4.45	54.2	20.660	+.613	52.4	47.7	4.7	68.0	E. 1	6	Nimbus stratus.	" "
	5.	53.4	20.658	+.611	52.3	47.2	5.1	66.0	E. 1	5	" "	" "
	5.30	53.0	20.656	+.611	50.6	46.2	4.4	71.6	E. 1	4	" "	" "
	6.	48.5	20.651	+.610	47.8	44.7	3.1	77.2	E. 1	3	" cirrus.	" "
	6.30	47.3	20.644	+.608	47.7	42.6	5.1	62.4	S. E. 1	2	Cirro stratus.	" "
	7.		20.644	+.610								

* Took barometer in. † Put barometer on tree again.

Barometer 2219, at Station in Camp Nassau. (4.)

DATE.	A.M.	T.	B.	32° B.	D. B.	W. B.	D —	R. H.	Wind, direction and force (1 — 10).	Clouds (1 — 10).	Nature of clouds.	REMARKS.
Saturday, August 18th, 1877.	7.	46.4	29.633	+.601	46.0	43.2	2.8	80.3	...	9	Cumulus nimbus.	Cool and cloudy.
	7.30	49.6	29.639	+.601	48.5	45.0	3.5	74.3		9	"	"
	8.	51.3	29.646	+.605	49.7	45.2	4.5	67.6	S. 1	9	"	"
	8.30	53.6	29.650	+.605	51.6	46.5	5.1	65.3	E. 1	8	"	"
	9.	53.4	29.654	+.609	51.6	46.8	5.8	65.3	S.E. 2	9	"	"
	9.15	53.9	29.655	+.610	52.1	46.7	5.4	61.8	S. 2	9	"	"
	9.30	53.4	29.656	+.611	50.6	45.1	5.5	65.3	S. 1	9	"	"
	9.45	53.6	29.659	+.614	51.4	45.9	5.5	61.8	S. 1	9	"	"
	10.	53.8	29.661	+.616	52.3	46.8	5.5	62.6	S. 1	9	"	"
	10.15	54.4	29.662	+.617	52.5	46.8	5.7	62.6	S. 1	8	"	"
	10.30	55.3	29.662	+.615	57.7	49.3	10.9	37.9	S.E. 1	6	Cumulus.	Sun out.
	10.45	58.8	29.660	+.611	56.7	47.6	7.4	56.9	S.E. 2	4	"	Pleasant.
	11.	59.2	29.660	+.606	58.3	51.1	7.9	54.5	S.E. 2	5	"	"
	11.15	60.7	29.658	+.602	58.3	49.8	7.2	58.4	S. 2	6	Nimbus.	Rain.
	11.30	59.0	29.658	+.600	59.5	49.8	6.2	56.9	S.W. 3	8	"	
	11.45	57.8	29.660	+.602	55.0	49.8		61.3		10	"	
	P.M. 12.15	56.4	29.656	+.608	53.5	50.1	3.4	79.8	S. 1	10	"	Stopped a little.
	12.30	57.3	29.658	+.666	54.1	49.2	4.9	70.2	S. 1	9	Cumulus nimbus.	Cloudy; rain.
	12.45	56.2	29.659	+.606	53.4	49.8	3.6	76.3	S. 1	9	"	Raining still.
	1.	53.6	29.658	+.609	51.2	48.6	2.8	82.2	W. 1	10	"	"
	1.15	52.4	29.653	+.613	51.1	47.5	3.4	78.6		10	Nimbus.	Rain stopped.
	1.30	51.8	29.650	+.610	51.4	46.8	3.6	75.3		10	"	Calm.
	1.45	52.5	29.648	+.607	51.9	47.0	4.6	68.3		10	"	"
	2.	54.7	29.646	+.603	53.2	47.2	4.9	68.9		10	"	"
	2.15	55.4	29.644	+.597	53.3	48.7	5.2	(6.0)		10	"	"
	2.30	53.4	29.630	+.591	52.3	47.2	4.5	69.6		9	"	"
	2.45	53.7	29.628	+.583	52.3	47.3	4.8	68.9		9	"	"
	3.	57.8	29.625	+.580	52.3	47.8	4.5	68.9		9	"	"
	3.15	51.7	29.625	+.580	51.0	47.4	3.6	75.3	E. 1	9	"	"
	3.30	51.8	29.629	+.586	50.8	47.2	3.7	75.3	S. 1	10	"	Cool.
	3.45	50.1	29.631	+.591	49.1	45.4	4.1	74.3	W. 1	10	"	"
	4.	50.6	29.630	+.591	49.4	45.3	4.5	71.0	W. 2	9	Cumulus	"
	4.15	51.5	29.631	+.591	49.6	45.1	4.2	67.6	W. 2	9	"	"
	4.30	51.8	29.633	+.592	49.2	45.0	3.7	71.0	W. 1	9	"	"
	4.45	50.4	29.630	+.591	48.7	46.2	3.6	74.3	S. 1	9	Cumulus stratus.	Rain.
	5.	51.2	29.628	+.587	49.8	46.0	2.7	74.8		7	"	"
	5.30	50.0	29.628	+.589	48.7	45.9	1.0	81.5		8	"	"
	6.				46.9			92.1	...	9		
	6.30	50.0	29.628	+.589	46.9	45.9	1.0	92.1				
	7.	49.8	29.628	+.590	46.8	45.6	1.2					

Barometer 2219, at Station in Camp Nassau. (4.)

DATE	A.M.	T.	B.	32° B.	D. B.	W. B.	D—	R. H.	Wind, direction and force (1—10)	Clouds (1—10)	Nature of clouds	REMARKS.
Monday, August 20th, 1877.	9.30	54.2	20.662	+.615	50.5	43.4	7.1	51.5	N. W. 3	4	Cumulus.	Cool and pleasant.
	9.45	54.5	20.662	+.615	52.0	44.0	8.0	47.2	N. W. 4	5	" and stratus.	" "
	10.15	56.2	20.662	+.612	55.5	44.3	11.2	33.3	N. W. 5	6	"	" cloudy.
	10.30	55.0	20.660	+.611	54.8	43.5	11.3	32.0	N. W. 6	7	"	" " "
	10.45	57.5	20.660	+.608	56.2	44.8	11.4	33.3	N. W. 6	8	"	" " "
	11.	59.0	20.660	+.604	57.5	45.0	12.5	28.6	N. 5	8	"	Sprinkling.
	11.15	57.5	20.658	+.606	55.0	43.5	11.5	29.8	N. 4	9	"	Little rain and hail.
	11.30	55.0	20.656	+.605	53.5	43.0	10.5	34.1	N. 5	8	"	Clearing.
	11.45	56.5	20.658	+.608	54.0	44.5	9.5	39.9	N. W. 6	7	"	" "
	12.	58.0	20.656	+.600	56.5	44.0	12.5	27.2	N. W. 3	6	"	Cool and cloudy.
P.M.	12.15	59.5	20.656	+.604	58.0	44.2	13.8	23.5	N. W. 4	7	"	" "
	12.30	57.0	20.656	+.606	55.5	43.0	12.5	25.8	N. W. 3	8	"	Rain.
	12.45	56.0	20.654	+.602	54.2	41.5	12.5	22.8	N. 5	7	"	" "
	1.	57.5	20.654	+.604	54.5	42.0	12.0	26.4	N. 6	9	"	Cool and clearing.
	1.15	57.8	20.654	+.598	55.0	42.5	12.5	25.8	N. 5	9	"	" " "
	1.30	59.5	20.654	+.605	57.0	43.0	12.5	27.2	N. W. 6	9	"	Hail.
	1.45	55.5	20.654	+.593	52.5	44.5	11.3	28.4	N. 7	8	"	Clearing.
	2.	53.5	20.650	+.601	54.5	42.8	11.7	46.6	N. 6	9	"	Cool.
	2.15	55.8	20.650	+.607	49.5	40.4	9.1	37.3	N. 7	8	"	Clearing.
	2.30	52.5	20.650	+.605	49.2	40.4	7.8	36.3	N. 6	8	"	Cloudy and cool.
	2.45	51.5	20.648	+.607	49.5	41.4	9.1	25.4	N. 7	7	"	Sprinkling.
	3.	54.5	20.648	+.601	53.8	43.8	10.0	24.9	N. W. 5	8	"	Cloudy.
	3.15	53.8	20.648	+.603	52.2	43.2	11.7	24.9	N. W. 6	8	"	Cool and cloudy.
	3.30	54.5	20.648	+.598	53.4	43.0	12.4	27.6	N. W. 6	7	"	Clear.
	3.45	56.5	20.648	+.601	53.8	41.0	12.4	32.2	N. W. 5	6	"	" and cool.
	4.	43.5	20.646	+.603	51.0	40.0	11.0	41.2	W. 4	6	"	" " "
	4.15	52.8	20.646	+.603	50.5	40.0	10.4	48.3	W. 5	5	"	" " "
	4.30	52.0	20.644	+.605	49.5	41.0	10.1	52.3	W. 5	4	"	" " "
	4.45	50.5	20.644	+.603	47.2	41.2	8.5	51.3	W. 3	4	"	" " "
	5.	51.0	20.644	+.606	48.0	40.4	7.2	54.3	W. 2	3	Cumulus.	" " "
	5.30	49.5	20.644	+.603	47.2	40.2	6.8	51.3	W. 4	5	"	" " "
	6.30	48.4	20.644	+.610	46.2	40.4	6.6	64.1	W. 3	3	"	" "
	7.	47.0	20.642	+.613	45.0	40.2	6.0	86.9	W. 1	2	"	" and cold.

Barometer 2219, at Station in Camp Nassau. (4.)

DATE	A.M.	T.	B.	32° B.	D. B.	W. B.	D –	R. H.	Wind, direction and force (1 – 10).	Clouds (1 – 10).	Nature of clouds.	REMARKS.
Tuesday, August 21st, 1877.	9.	53.8	20.642	+597	52.2	39.8	12.4	21.6	N. W. 5	4	Cumulus.	Cool and cloudy.
	9.15	53.5	20.642	+597	52.0	40.0	12.0	23.3	N. W. 4	5	"	" "
	9.30	55.2	20.642	+593	54.5	42.0	12.5	24.3	W. 4	6	"	" "
	9.45	54.5	20.642	+595	52.2	40.0	12.2	23.3	W. 5	6	"	" "
	10.	55.0	20.642	+593	53.0	40.5	12.5	21.2	W. 4	6	"	" "
	10.15	....	....	....	....	....	....	....	....	:	.....	.....
	10.30	....	....	....	....	....	....	....	....	:		:
	10.45	57.5	20.642	+590	56.8	41.2	15.6	11.8	W. 5	7	Cumulus and str.	Hail.
	11.	56.2	20.642	+592	52.5	40.8	11.7	25.4	W. 6	9	"	"
	11.15	55.0	20.640	+591	52.0	41.0	11.0	29.2	W. 6	9	"	
	11.30	....	....	....	....	....	....	....	....	:		:
	11.45	....	....	....	....	....	....	....	....	:		:
	12.	57.4	20.642	+599	54.5	41.4	13.1	20.8	N. W. 6	8	Cumulus and str.	Clear.
	P.M.											
	12.15	58.0	20.642	+588	56.2	42.0	14.2	18.6	N. W. 5	7	"	Cloudy.
	12.30	58.5	20.644	+588	56.0	42.5	13.5	20.5	N. W. 4	7	"	Clear.
	12.45	59.2	20.644	+588	55.5	42.0	13.5	20.5	N. W. 5	6	"	" "
	1.	60.0	20.644	+586	56.0	42.0	14.0	18.6	N. 5	7	"	" "
	1.15	61.5	20.644	+584	57.0	42.2	14.8	16.8	W. 6	8	"	Cloudy.
	1.30	61.6	20.644	+584	58.0	42.8	16.2	11.9	W. 6	8	"	"
	1.45	62.0	20.640	+584	57.5	43.0	14.5	18.3	W. 5	9	"	Clear.
	2.	62.5	20.636	+579	56.8	42.8	14.0	18.6	W. 4	7	Cumulus.	Cloudy.
	2.15	60.0	20.632	+577	55.0	42.5	12.5	24.3	W. 7	8	"	"
	2.30	55.8	20.630	+582	54.5	40.2	14.2	15.4	W. 7	8	"	Clear.
	2.45	58.2	20.630	+577	57.5	42.4	15.1	15.1	W. 5	7	"	Cloudy.
	3.	58.6	20.630	+572	54.0	40.5	13.5	17.2	W. 5	8	"	"
	3.15	56.8	20.626	+581	56.5	42.8	13.7	20.5	W. 6	7	"	Clear.
	3.30	54.5	20.624	+576	56.0	42.2	14.8	16.8	W. 5	6	"	"
	3.45	56.2	20.622	+576	55.0	40.8	13.8	20.5	W. 4	5	"	Clear.
	4.	54.5	20.622	+576	53.5	40.0	14.2	15.4	W. 3	4	"	" and cool.
	4.15	54.5	20.622	+572	54.8	42.0	13.8	17.2	W. 5	4	"	" "
	4.30	52.5	20.620	+573	53.8	39.5	12.8	24.3	W. 5	4	"	" "
	4.45	53.5	20.620	+575	53.2	39.0	14.3	14.3	W. 5	3	"	" "
	5.	48.2	20.620	+577	50.5	38.0	14.2	13.7	W. 6	2	"	cold.
	5.30	45.4	20.618	+579	50.0	37.5	13.5	13.7	W. 6	2	"	" "
	6.30		20.614	+582	47.0	36.8	12.5	16.0	W. 2	1		" "
	7.			+584	45.0	35.4	10.2	26.0	W. 4	0	.....	" "
							9.6	27.1		0		" "

Barometer 2219, at Station in Camp Nassau. (4.)

Date	A.M.	T.	B.	32° B.	D.B.	W.B.	D —	R.H.	Wind, direction and force (1 – 10).	Clouds (1 – 10).	Nature of clouds.	Remarks.
Wednesday, August 22d, 1877.	9.	54.5	20.618	+.571	51.5	38.0	13.5	13.7	S. W. 4	1	Stratus.	Clear and cool.
	9.15	53.8	20.618	+.573	51.8	38.0	13.8	13.7	S. W. 4	1	"	"
	9.30	54.2	20.618	+.571	53.8	38.6	14.2	11.9		0		"
	9.45	55.5	20.620	+.571	53.0	39.2	13.8	15.5	S. W. 5	0		"
	10.	55.8	20.618	+.569	54.5	40.0	14.3	15.4	W. 4	0		and pleasant.
	10.15	56.6	20.618	+.568	54.2	40.2	14.2	15.4	W. 3	0		"
	10.30	57.2	20.618	+.566	55.5	41.0	14.8	13.6	W. 5	0		"
	10.45	57.8	20.618	+.566	55.0	40.6	14.5	15.3	W. 5	0		"
	11.	58.0	20.620	+.566	55.5	40.8	14.4	15.4	S. W. 5	0		"
	11.13	58.0	20.620	+.566	56.2	41.0	14.7	13.6	S. W. 4	0		"
	11.30	59.0	20.620	+.564	57.4	41.5	15.2	11.8	S. W. 4	0		"
	11.45	60.6	20.620	+.562	56.6	43.4	15.9	23.9	S. W. 5	0		"
	12.	60.0	20.620	+.562	13.2							"
P.M.												
	12.15	60.8	20.620	+.562	56.0	42.2	13.8	20.5	W. 3	0		"
	12.30	61.4	20.620	+.560	59.0	41.4	17.6	5.7	W. 4	0		"
	12.45	60.5	20.616	+.558	58.2	41.0	18.2	2.5	W. 3	0		"
	1.	61.5	20.616	+.556	54.6	40.0	13.6	18.9	W. 4	0		"
	1.15	62.2	20.616	+.555	58.2	40.0	18.2	2.5	W. 5	0		"
	1.30	62.5	20.616	+.555	59.0	41.2	17.8	5.7	W. 4	0		"
	1.45	63.0	20.616	+.555	60.0	42.2	17.8	7.4	N. W. 3	0		"
	2.	64.2	20.616	+.553	60.5	42.0	18.2	6.0	W. 4	0		"
	2.15	64.5	20.616	+.551	62.5	42.2	18.3		W. 5	0		"
	2.30	64.0	20.612	+.557	62.0	41.8	20.5		W. 3	0		"
	2.45	63.8	20.612	+.557	61.0	42.2	20.2	4.6	W. 4	0		"
	3.	61.8	20.610	+.546	62.0	42.0	18.8	6.0	W. 6	0		"
	3.15	61.5	20.606	+.544	60.0	42.2	18.0	8.8	W. 4	0		"
	3.30	62.0	20.604	+.541	60.0	42.5	17.3	7.4	W. 5	0		"
	3.45	61.8	20.602	+.542	59.8	42.4	17.5	7.4	W. 4	0		"
	4.	61.5	20.602	+.540	59.5	42.2	17.6	7.4	W. 3	0		"
	4.15	60.0	20.600	+.542	59.0	42.0	17.5	7.2	W. 4	0		"
	4.30	59.8	20.600	+.544	57.0	41.8	17.2	6.9	W. 5	0		"
	4.45	58.5	20.596	+.544	56.0	40.2	16.2	6.7	W. 6	0		"
	5.30	57.0	20.588	+.542	53.2	39.8	16.2	10.1	W. 4	0		Cool.
	6.	54.0	20.584	+.537		38.5	14.7		W. 3	0		Clear.
	7.	46.0	20.576	+.544	42.5	37.0	15.5	4.6	W. 4	0		"

Barometer 2219, at Station in Camp Nassau. (4)

DATE.	A.M.	T.	B.	32° B.	D. B.	W. B.	D –	R. H.	Wind, direction and force (1 – 10).	Clouds (1 – 10).	Nature of clouds.	REMARKS.
Thursday, August 23d, 1877.	9.	60·8	20·616	+·558	59·0	41·5	17·5	5·7	W. 3	2	Stratus.	Clear.
	9.15	60·5	20·618	·560	58·5	41·2	17·3	7·2	W. 2	2	"	"
	9.30	61·8	20·622	·562	60·5	42·8	17·7	7·4	S.W. 3	2	"	"
	9.45	62·0	20·624	·564	60·2	42·2	18·0	6·0	S.W. 2	2	"	"
	10.	62·5	20·624	·563	60·3	42·4	17·8	7·4	S.W. 2	1		"
	10.15	63·0	20·624	·563	61·0	42·8	18·2	6·0	S.W. 4	0		"
	10.30	64·0	20·624	·561	62·4	43·0	19·4	5·0	S.W. 3	0		"
	10.45	64·5	20·626	·559	63·2	43·8	19·4	5·0	S.W. 3	0		"
	11.	64·0	20·626	·559	63·0	43·2	19·8	3·7	S. 3	2		"
	11.15	64·5	20·626	·561	62·4	44·0	18·4	9·2	W. 3	2		"
	11.30	65·5	20·626	·561	64·4	44·5	17·9	10·6	W. 4	3		"
	11.45	67·0	20·626	·559	64·0	44·5	20·0	4·3	W. 4	5		"
	12. P.M.	67·5	20·626	·555	65·8	41·0	24·8	...	W. 3	6	Cumulus.	Cloudy.
	12.15	66·0	20·626	·555	66·0	43·8	22·2	...	S.W. 4	7	"	"
	12.30	65·2	20·626	·559	62·2	43·8	18·4	7·6	S.S.W. 5	8	"	"
	12.45	64·8	20·628	·561	61·8	42·6	18·9	4·6	S.S.W. 4	8	"	"
	1.	65·0	20·628	·563	61·5	42·4	19·4	3·3	S.W. 4	8	"	"
	1.15	65·2	20·624	·557	61·5	43·0	18·5	2·1	S.W. 4	8	"	"
	1.30	64·5	20·624	·557	62·5	43·0	18·5	6·3	S.W. 5	8	"	"
	1.45	63·0	20·624	·559	60·8	42·6	19·3	5·0	S.S.W. 3	8	"	"
	2.	64·6	20·624	·561	60·2	42·5	18·2	6·0	S.S.W. 4	7	and nimbus.	Clear,
	2.15	63·0	20·624	·559	61·5	43·0	17·7	4·7	S.S.W. 5	6	"	"
	2.30	64·6	20·620	·559	65·5	43·0	18·5	6·3	S.S.W. 4	5	"	Cloudy.
	2.45	61·2	20·620	·555	62·0	42·8	21·5	...	S. 4	4	"	Clear.
	3.	64·0	20·618	·551	63·5	44·0	19·2	3·3	S.W. 2	7	"	Cool.
	3.15	65·5	20·608	·547	60·0	41·5	19·5	3·7	S. W. 1	6	"	"
	3.30	62·3	20·608	·547	59·6	41·5	18·5	3·0	S.W. 1	6	"	"
	3.45	62·5	20·600	·542	59·1	40·5	18·6	4·3	S.S. 2	6	"	"
	4.	60·3	20·599	·541	58·8	41·7	18·6	7·2	W. 2	9	"	"
	4.15	60·1	20·600	·542	58·8	40·8	17·1	2·5	W. 3	10	"	"
	4.30	59·5	20·602	·546	58·4	41·1	18·0	7·2	W. 2	10	"	"
	4.45	59·3	20·600	·544	57·6	39·8	17·3	2·1	S. 3	10	"	"
	5.	59·6	20·598	·542	58·1	40·5	17·8	3·9	S.E. 2	9	"	"
	5.30	59·2	20·595	·539	58·4	39·8	17·6	...	S. 2	7	stratus.	"
	6.30	56·8	20·580	·539	55·5	39·6	15·9	8·4		4	"	"
	7.	55·8	20·584	·535	56·3	38·6	15·7	6·5		3		"

APPENDIX.

DIARY OF THE TRIP

.

DIARY OF THE TRIP.

AT 7 P.M. the members of the expedition assembled in the study of Prof. Kargé, and thence proceeded to Dr. Mc-Cosh's house, to bid him good-by. He called us into his study and prayed for our safety and success, and bid us all an affectionate good-by. We then went on the same errand to Dr. Guyot's, where he and his good wife wished us all possible success, and hoped we should bring back our own bones in safety, as well as many fossils. When we reached the depot we found a large number of the students had gathered, and ' they sent us off with the old-time cheer and rocket. Leaving there at 8.10 we arrived at the Junction, where we said good-by to many of our friends and waited for the 10.20 train which was to stop and attach our baggage-car and Pullman, which were waiting there for us. Our Pullman car was the best on the road ; it was the President's, and was kindly placed at our disposal by Mr. Pullman. Its name was the Rhode Island. Our berths were portioned out by lot, and we had No. 11, which we soon turned into, and with the exception of being waked up at Philadelphia by some fellows who boarded the train there, passed a quiet night.

We all awoke early, as many of us had not seen the Horse Shoe curve, and those who had were almost as anxious to see it again. Reached Pittsburg at two o'clock and took dinner. Then going over the Pittsburg, Fort Wayne and Chicago Railroad through Ohio, nothing of interest occurred till we reached Alliance. We had now for the most part assumed our blue shirts and uniform costumes, which attracted the attention of the natives of that rustic town. One of their number, with more valor than discretion, singling out General

4

Kargé as the leader of the party, approached him, and in the mildest tones asked, " Is this a travelling base-ball club, sir?" He was immediately squelched by the stentorian tones of the General, as he replied, "No, sir; this is a circus!" and the General was not far from right, for the whole company afterward proved a source of entertainment to many a native of the far West.

JUNE 23, SATURDAY.

We arrived early this morning in Chicago, where we were met at the depot by Mr. Hathaway, of the Grand Pacific Hotel, who kindly invited us to partake of a very elegant breakfast given as a compliment to our party. We took our chronometer to the Douglass Observatory and compared it with the standard there, and at 12.30 met the rest of the party (who had separated to visit various points of interest in the city) at the Chicago and Alton depot. We took possession of the palace-car " Baltic," and giving three cheers and a rocket for our friends in Chicago, who had gathered at the depot to see us off, we started on our long prairie ride. At twelve o'clock in the evening, as we were approaching the bridge at Louisiana, across the Mississippi, a number of us were enjoying the moonlight in the engine-cab, when we saw a sight which curdled our blood. A trackman who had just been paid off was going home intoxicated, and had lain down on the track to sleep. Before we could say a word we were on and over him. We stopped, went back, and found his body, and did what we could for him, but to no avail; so we left him at a section-house to die.

JUNE 24, SUNDAY.

The first rays of light found us rolling along the muddy bottom of the Missouri. The unanimous verdict concerning our breakfast at Lexington Junction was that it was "very poor." Soon after leaving this place we encountered a novel experience in railroading. We plunged in water up to the top of the wheels, and moved along at a snail's pace for about a mile and a half, constantly expecting to be stopped by the fires being put out in the engine. We came into Kansas City two hours late. Here we found, to our astonishment, that our baggage-car would not fit the rails of the Kansas Pacific Road, and our whole party had to turn in and transfer our baggage from one car to another. We did this in the short space of twenty minutes, though we had upwards of five thousand pounds to move. This gave quite a reputation to our "tender-foot" party for muscle and energy. The most of the afternoon we spent in singing Moody and Sankey's hymns, with Dunning as leader of the choir.

JUNE 25, MONDAY.

We awoke to find ourselves rolling out toward the plains. We spent the morning in the baggage-car, where some of us tried our marksmanship on the antelope we passed ; none of them, however, suffered. About noon we passed a small station called First View, one hundred and sixty miles from Pike's Peak, and as an illustration of the clearness of the atmosphere, we would mention the fact that the tip of the snow-capped mountain was just in sight over the rolling prairies. We reached Denver at six in the evening, and after investigating the town, took supper at Charpiot's Restaurant, "the Delmonico (?) of the West."

JUNE 26, TUESDAY.

We pitched our camp a half mile west of Denver, and commenced assorting our baggage for our proposed trip in Colorado. This was our first regular camp and our first experience under military discipline. The camp was called "Camp Lynch," because we discovered that three horse-thieves had been lynched there some years before.

JUNE 27, WEDNESDAY.

We spent all day in camp selecting horses. As the animals from which the party were to make a selection were Indian ponies, lately captured from the Sioux tribe, the efforts on the part of some of the members of the party to ride them caused considerable amusement, and their motions were more grotesque than graceful.

JUNE 28, THURSDAY.

At four in the afternoon, everything being in readiness, we broke camp; and falling into proper marching order, we passed through the city to the astonishment of its peaceful inhabitants, who flocked out to see the warlike display. We took up our line of march along the South Fork of the Platte River, and camped four miles out of the city on its banks. Dulles and Dunning mounted guard for the night.

6

JUNE 29, FRIDAY.

Continuing on in the same line southward, after passing
Littleton, we turned to the left along Plum Creek for thirteen
miles, and then camped at Van Wormer's ranche. Here the
party indulged in the home luxuries of milk and eggs to an
enormous extent.

JUNE 30, SATURDAY.

When we left the camp this morning we received the
cheering news that military discipline was at an end, and that
each one might direct his horse where his inclination led him,
provided he came to the appointed rendezvous at night.
Leaving the main party, we struck out straight across the
foot-hills for Castle Rock, so named from a square rock,
about 90 feet high, on the top of a mountain overlooking the
railroad station. As we rode into the village we saw a sign
with the enticing words, "Strawberries and ice-cream," on it,
and forthwith prepared ourselves to enjoy them. But on a
closer inspection of the aforesaid sign, we found printed below, in microscopic characters, "On the 4th"; so we had to
be satisfied with some sour strawberries and some very dry
ginger-snaps. We went on about eight miles further to Leverson's ranche, the appointed meeting-place for the night.

Here we waited for the rest of the party, who came straggling in one by one, and finally Lynde brought the startling
information that the "commissary department" had broken
down six miles back; and as the housekeeper was not prepared for the reception of sixteen hungry troopers, the chance
of our getting a supper seemed to be rather slight. Finally,
not caring to have us starve, she gave us freely of what she
had in large store, viz., milk and biscuits. She also very
kindly offered us for sleeping purposes the only empty room
she had in the small log cabin. Taking our saddles for pillows,
and fully booted and spurred, we laid down; but not to
peaceful slumbers.

JULY 1, SUNDAY.

Sabbath morning was spent in anxiously awaiting the arrival of the wagons. They finally reached us at 12 M. All
hands then set to work to prepare dinner, which was to
answer for two meals. We enjoyed this meal more than any
that we had since starting out. The afternoon was spent

quietly in camp; the duties of correspondence were faithfully attended to; and the kind hospitality of Dr. Leverson was very refreshing and fully appreciated.

JULY 2, MONDAY.

After leaving Leverson's ranche we soon reached Larkspur, and, turning to the right, passed across the plains and came to Lake Divide, which is on the summit of the transverse . range of hills separating the Platte River from the Arkansas River. The elevation is 7,554 feet. Five miles from the lake we came to Monument, and encamped about a mile further on, near the road. We took a tramp up the Little Giant mountain, from which we had a beautiful view of Pike's Peak and the Front Range. These sights were new and strange, and they filled us with admiration for the sublime nature we were brought in such close contact with. On our road up and on the return we passed through long, beautiful meadows with high eroded columns of sandstone, which seemed very grand. The whole seemed to us rather more like an enchanted land than a reality.

JULY 3, TUESDAY.

Starting early in the morning, and passing through Edgerton, we skirted along the eastern edge of Monument Park. We reached Colorado Springs at twelve o'clock, having been on the road four days and a half, and having ridden seventy-five miles. After a good dinner we set out again to find our party, and rode over to Glen Eyre, which was full of natural wonders new to us, in its curiously upturned strata. Not finding them there, we went to the Garden of the gods, where the wagons were already drawn up, and a permanent camp formed near the Gateway.

JULY 4, WEDNESDAY.

The members of the various departments represented in the expedition now set to work in real earnest; and we commenced our part by establishing the height of a base at the Beebee House in Manitou Springs for the purpose of measuring Pike's Peak. This was finished by noon; and M. choosing the lower station, L. started, after dinner, to climb the mountain, and reached the Lake House in the evening after some adventures not worth noting.

8

L. started early for the summit, having agreed with M. to carry on simultaneous observations from 9.30 to 10.30 A.M. This was accomplished, and our first attempt at measuring these mountains will be seen to have been successful and very encouraging. The mountain was descended with ease, and the camp reached by evening. The incidents of the climb were so like those of other parties performing the same feat, that they are hardly worth repetition.

In order that we might ascertain the elevation of our base at the Beebee House, a series of observations was conducted in the morning between L. at that place and M. at the Colorado Springs Railroad depot. The afternoon was spent arranging our instruments in the camp.

We spent the day visiting the Springs and sampling them all. They were chiefly soda, iron, and magnesia. We took the temperature of the various springs, and found it to vary from 58° to 68° F.

We spent most of the day dodging the rain, as it poured hard all the time. Attended the only Episcopal service which was held in the morning.

This morning we left the main party in their camp and went on ahead in order to be able to prosecute our work. Leaving Manitou at 9.30 we commenced our climb over the Front Range by the way of the Ute Pass. The scenery was very beautiful, and was enjoyed as much as possible as we hurried along on our ponies in light travelling order. At two o'clock we reached " Silver Spring Ranche," having passed a very pretentious hotel called the " Green Mountain House," where we were informed by the landlady that she " had quit keeping stoppers." The proprietor of the Silver Spring Ranche was the far-famed Dr. Johnson, and he treated us very nicely. After dinner we rode fifteen miles farther on to Florissant, where we were entertained in royal style by the stately host of the place, Judge Castello. This old place was crowded with relics of the chase and curiosities of all sorts. There was an old Indian fort directly back of the house; it had a circular wall and several outworks. We examined them all carefully, and in the evening the Judge regaled us with stories about old times, and his encounter with the Indians. (We have since learned of the death of our good old friend, with regret.)

Four miles from Florissant we crossed the north branch of the South Fork of the Platte River. From here for seven miles our road lay through a beautiful district leading to South Park—high granite walls and immense boulders lying on either side of the way. Coming to Wilkinson's ranche we were refused water for our horses and had to ride on. After passing over a small divide we saw the broad expanse of South Park some hundreds of feet below, stretched out like a large map, dotted here and there with a ranche, and the long Platte River shining like a silver line throughout its entire length. While crossing this open piece of prairie to what appeared to be the nearest ranche, we saw a most wonderful display of clouds. The heavy cumulus masses rose up higher and higher till they rolled over one bank after another in great confusion and presented a most imposing sight. Reaching the ranche, we approached the house, hoping to get some water, but the door was slammed in our faces by a wrathful female, who screamed out that she "didn't have no water for stoppers." Our feelings at such a reception can be imagined better than expressed. We next went over to Hartsell's, where we at last found what we wanted, after having come for it nineteen miles over a very hot road. But here was another difficulty; a "round-up" was in progress, and all available quarters were occupied, so we had to urge our way on again. A round-up is the technical term out here for a gathering of all the cattle on the park, for the purpose of branding the calves and counting the herds. This is held every year. Large numbers of cattle were already there, and many more were to come. This was the favorite spot for this purpose, as water and grass are very plenty. Finally we reached Clark's, well tired out and rejoiced at the prospect of a rest. We put our horses up, and went off to take a look at the neighborhood, which we found very pleasant.

5

In the evening we experimented practically on the benefits and efficacy of the hot sulphur spring near the house, and were delighted. Its temperature was 150° F.

Leaving Clark's, we soon came to a division in the road, which, a sign told us, was the way to the much-sought-for San Juan country. The road looked well worn. We had frequently passed, since leaving Colorado Springs, the teams of travellers who were slowly going to and from that Mecca of the gold-hunter. We reached Fairplay at eleven ; and M. staying on the site of an old building whose elevation was known, L. went on to Alma in order to make a station at that place, preparatory to measuring several mountains in that region. This work occupied the rest of the day. The town of Alma is considered the highest mountain town in the State, being, according to our measurement, 10,381.5 feet at the St. Nicholas Hotel. Our barometer stood at 20.8 inches, with an average temperature of 72° F. Fairplay is noted for its gulch-mining interests, and the hillsides are tunnelled for miles in all directions by the eager miners. We noticed here the cheerful faces of many Chinamen hard at work. The process is to empty the gold-bearing earth into long sluices, through which runs a rapid stream of water strong enough to carry large boulders, but not powerful enough to wash away the gold, which sinks to the bottom, where it is caught by the mercury behind the riffle-bars, which· are placed at intervals transversely to the current. This amalgam is taken out and the gold obtained by volatilizing the mercury. The "wash," as it is called, averages here $11 a day per man.

We made corresponding observations in the morning until nine o'clock, when L. started for Mount Lincoln, having previously procured a horse and guide. M. removed his instrument to the spot that had been left by L., and, as agreed upon, continued observations there. On the road up Mount Lincoln, on the sides of Mount Bross, L. visited some mines rich in silver ore. He reached the top of Mount Bross at 10.30, made observations for half an hour, and then continued on the road up Mount Lincoln, the summit of which was reached at 12. Here observations were made; and after taking dinner at some miners' quarters near the top, he commenced the descent. Before reaching the bottom, however, he was quite overcome by the "mountain fever," as it is called there. It is a mild form of typhoid fever, and very harmless if taken care of in time.

JULY 13, FRIDAY.

M. started at 7.30 A.M. for Mount Quandary with the horse and guide, L. remaining in Alma to make corresponding observations, as he felt much better to-day. M. rode up to the entrance of the once famous Montgomery Gulch (where so many thousands of dollars were sunk in 1863, when Montgomery was the largest town in the State; now nothing remains but the ruins of the place). Thence going to the summit of the Hoosier Pass, he turned to the left, and after a very hard and rough climb he reached the top at 12.30, where observations were made. An interesting monument was found on the summit, containing a board, on which was carved, "Kelsey and Tom Campbell, July, 1873." Thence he returned to Alma, reaching it at 6 P.M.

JULY 14, SATURDAY.

M. started to measure Mount Silverheels, leaving L. below. This mountain is a prominent point on the Park Range, affording a good view of the Platte River and South Park. This would be a fine place for transit observation if it were not for the iron present in the mountain, which rendered a small compass completely useless. In the evening the town was enlivened by a horse-race in the streets. All the gambling saloons were open (and it seemed as if almost every other house was one), and the drinking and carousing were carried on till late. Alma has one of the largest hydraulic mines and reduction mills in the State. The principle of these mines is the same as the gulch mines, except that the source of the power is a strong head of water delivered from a large nozzle against the bank of earth, which it undermines and makes short work with.

JULY 15, SUNDAY.

The scenes of the night before were continued all day. Everything went on as usual, except where mines had closed for a wash-up. Sunday is a day for general dissipation, and is spent by the miners in the wildest sorts of vice. Churches are unknown, and there is nothing to remind any one that it is not a week-day. M. went to Fairplay, as we had heard that Osborn was there, with a message from the main party. He came back to say that they intended to go to the Arkansas Valley, and wanted us to come with them. In the evening two men were shot in front of the hotel, in a little row that took place.

JULY 16, MONDAY.

We left Alma at 9 A.M., and found McPherson at Fairplay. We proceeded to the salt works, twenty miles south, where we got our dinner. The rides over the prairie land had a general sameness, which became very tiresome. These salt works are now abandoned as unprofitable. Leaving here, we started in full pursuit of the party, who we learned had spent Sunday here, and had just gone on. This part of the day's journey was very pleasant, as the Trout Creek Valley is quite a pretty place. We overtook the party just as they were going into camp in the Arkansas Valley, after a ride of sixty miles, the longest as yet made on the trip.

JULY 17, TUESDAY.

M. rode down the valley of the Arkansas six miles to Helena, the height of which place was known, and prepared to make corresponding observations; while L. set off early for Mount Princeton, a very beautiful and symmetric mountain immediately to the west of our camp. L. found no particular difficulty until within 1,500 feet of the top, when his only way lay over a bed of *débris* such as he had never experienced before; the size of the boulders being such that nothing but the hardest sort of crawling would answer. The summit was, however, reached at 12.30 P.M., and several patriotic demonstrations gone through with. A large monument, six feet high by three in diameter, was built, in the centre of which a stick, on which the circumstances were written, was deposited and left as a memento. The scene from this point of view was simply indescribable; the broad valley of the Arkansas spreading immediately below, and the Sawatch Mountains stretching away on either side. It rivals in beauty the scene from Mount Lincoln. It does not combine such an extent of prairie with great mountain regions as does Pike's Peak, but for pure mountain scenery it is very grand. L. returned to camp in the evening to find all the rest gone out fishing, as the rumors of the excellent trout to be found in the neighboring streams were not to be disregarded until verified by even the most unscientifically inclined; 101 of the speckled beauties were caught. M. caught three that weighed nine pounds, one of them weighing three and a half pounds. The principal trouble seemed to be the scarcity of bait, as there was more trouble catching grasshoppers than fish. Every one had fish for supper, and was happy.

JULY 18, WEDNESDAY.

Leaving camp at 9 A.M., we went on ten miles to Riverside, where we met Scott, Speir, and Osborn, who had left the camp some days before on a fossil hunt. At this part of the ride we entered the narrow part of the valley. The rocky gorge down which the foaming river ran presented many varied and picturesque aspects. We soon reached Granite, another lifeless town. Getting our dinner there, we turned to the left to visit the Twin Lake region. On our way up we forded Lake Creek, where the water reached our horses' flanks. At the Lakes we found a very primitive hotel with cotton cloth partitions between the rooms, and a roof made of the tins of tomato cans; but it covered a very obliging and good-hearted hostess, who did all in her power to make it pleasant for us. These lakes lie at the foot of Mount Elbert, and each of them covers about 200 acres. They are plentifully supplied with an excellent variety of mountain trout, on which we tried our skill with some success.

JULY 19, THURSDAY.

The various members of the party came straggling into this camp during the day, and it was spent in various ways, hunting and fishing. A council was held, at which it was decided that General Kargé, with the Palæontologists, and ourselves, should go to Fort Bridger in Wyoming Territory; the Palæontologists to busy themselves in the neighboring bad lands, and we to go into the Uintah Mountains for the purpose of doing original topographical work.

Accompanied by Osborn, we left in the morning, and re-traced our way to Granite. Leaving after dinner we fol-lowed the Hayden trail straight over the mountains. After a very rough ride, in which we came near losing our way from the scarcity of trail-marks, we finally emerged into the trail leading over the Weston Pass. Following this down the mountain, we soon were out on the level tract of South Park on our way for Fairplay. We reached there at 9 P.M., and, much to our surprise, found that our afternoon ride had been thirty-eight miles instead of eighteen, as we had expected.

We travelled over the now familiar road between Fairplay and Alma, and just before we came to Hoosier Pass we were overtaken by a worn-out and dilapidated-looking individual who entertained us by his interesting reminiscences of fron-tier life. He led us over Fremont's trail of 1853, instead of the regular pass road, and made the trip to Breckenridge very pleasant by his company. Here we discovered that our fellow-traveller was the celebrated " Dick Allen," of the Fairplay *Sentinel*. After leaving the summit of the pass we descended into the valley of the Blue River, which we fol-lowed to the little town of Breckenridge, twenty-six miles from Fairplay. The first thing we saw was the grim face of Gen. Logan on the piazza, where he was surrounded by a crowd of admirers. Judge Silverthorn, who was the owner of this hotel, said that the house was full, but if we could sleep in the " corral" we might stay. We examined it and accepted the offer. It was in the upper story of the establishment, under the roof. While we were ridding ourselves of our numerous packages, the Judge disappeared. We found on going down that a drunken fellow had been behaving un-seemly in the street, and that our little shrivelled-up judicial authority had made out a warrant, served it himself, and seiz-ing the prisoner by the back of the neck, had kicked him up the side of a hill to the calaboose, although he had complained of being very ill a few minutes before. This little affair gave rise to a trial, which of course we attended. Judge Rieland read from a large and formidable-looking document, " Gibbs, you are charged with disturbing the peace." (*Aside*) "I'm in a hurry, so plead guilty. If you take one of them lawyer fellows, I'll stick you." Gibbs asked how much the fine was, and the Judge announced very formally that he should be fined $1.50 and costs. Whereupon Gibbs asked, in tones that made several."visitors" look around to see where the door was, " Where's my revolver?" He only wanted to pawn it, however. After this the court adjourned for drinks all around.

In the afternoon we went to see a wash-up in one of the neighboring gulch mines. Dick Allen went with us and explained the *modus operandi*. They took out $2800 for 160 days' work. When we returned home we found that there were six of us compelled to sleep in the seven-by-nine corral —the Judge, Dick Allen, we three, and a native hoosier.

JULY 22, SUNDAY.

We were awakened early this morning by Dick Allen saying, " I can't eat my breakfast until I have had six cocktails, and it's about time to commence; so come on, boys." We have spent some very queer days out here, but this beats them all. To get out of the reach of the noise was impossible, and you might think that there was a den of wild animals being fed, or something worse. We heartily recommend Breckenridge as being the most fiendish place we ever wish to see. We were forced to spend the morning and afternoon in the company of men whose language was vile, and whose actions were tinged with a shade of crime that shocked and hurt our senses ; never did anything so bestial and so unworthy even a mention by manly lips happen before our eyes. In the morning there was a series of dog-fights. A ring of yelling demoniacs was gathered around the two poor curs. Upon the success or defeat of one or the other large sums of gold-dust were freely wagered. The old Judge innocently remarked that " the men would soon commence," and sure enough they did commence very quick, quarrelling about the slightest thing, and fighting like devils while their breath lasted, which, on account of the rarity of the air at this altitude, was not more than ten minutes. Finally Breckenridge seemed satisfied, and grew quiet outside in the streets, but the men only adjourned to the saloons (almost every other house was one), where their carousing was kept up till late in the night. The evening being very cool, a fire was kindled in the parlor, around which a good many of the " old cases" were gathered, and the amusing stories of frontier and prairie experience served to pass away the hours till bedtime.

JULY 23, MONDAY.

Our friends gathered to see us off. Dick gave us a "send-off" in the way of a cock-tail, as his breakfast-time had not yet come. We were off at seven, and had a delightful ride for ten miles in the cool morning, down the valley of the Blue River. When we came to the Snake Valley, we rode up that stream ten more miles, through a most glorious piece of woodland scenery. We found we were in the neighborhood of Montezuma, and halted at the hotel for our dinner, which turned out to be a regular New York feast—mock-turtle soup, steak with mushrooms, etc. This fact, as well as the presence of New York papers, *Harper's Weekly* and *Monthly*, and the genteel appearance of our host, led us to make inquiries, whereupon he was recognized by two of us as one of those voluntary exiles from society, where he once moved with honor. Leaving at one o'clock, and riding five miles further up, we came to the head-waters of the Snake River at timber line, and saw our trail ahead, stretched along the precipitous side of the mountain which forms the head of the valley. Before and above us, almost 3000 feet, lay the ridge of the Argentine Pass, 13,100 feet in elevation, the highest wagon-road in the United States. After a long, slow ride, we reached the top, going sometimes over snowbanks (for it had not entirely disappeared here yet), sometimes finding barely walking room for our horses. From the summit the scene was perfectly grand. Here we stopped for our lunch, which we had in our saddle-bags, and rested our horses, while we enjoyed the scenery. A further ride of eight miles brought us to the hillside which seems to overhang Georgetown. This is a beautifully situated place, at the head of a valley surrounded by mountains on three sides. We stopped

17

at the Barton House, and received here the news of the great railroad riots in Pittsburg.

JULY 24, TUESDAY.

Before breakfast M. carried a series of levels from the depot of the Clear Creek Railroad up to the hotel, while L. compared the three barometers, as we were to leave an aneroid here with Osborn, and take the two mercurials respectively to Gray and Evans Peaks. M. started for Gray and found a good road to the top. This mountain is visited by many excursionists, being very accessible, and also being one of the highest in the range. L. took the road up Mount Evans, but soon reached a riding limit, and had to climb some 1500 feet before reaching the top. By means of a *powerful telescope he found that both observers were at work at the same time, which was very satisfactory. Storms of hail, snow, and rain were experienced by both, and the rain continued the rest of the day. Our canvas coats did good service on this occasion. After our return to Georgetown, and a little supper, we saddled our own horses (which were perfectly fresh, as we had left them here all day to rest) and started down the valley in the rain. We were already wet, and did not care much for a little more, as we wanted to gain on time and space. We raced all the way to Idaho Springs, making the fourteen miles in an hour and a quarter. We went to the Beebee House, which is kept by Mr. Beebee, who is not, however, as good at it as his wife in Manitou Springs. We were treated very nicely though ; had a cottage to ourselves, a wood fire, a good supper, and luxuriated in grand style.

JULY 25, WEDNESDAY.

We went to the Springs before breakfast, and enjoyed a warm soda bath very much. Leaving about nine o'clock we rode down part of the cañon, then turning to the right we went over St. George's Hill, and after this one hill was succeeded by another as we descended the foot-slopes of the mountains to the plains. We reached Mount Vernon at 1.30, where we had such an objectionable dinner set before us that its mere appearance disgusted us, and we rode on. It seemed good once more to be on the plains, after our four weeks in the mountains. Osborn left us here to see Prof. Lakes, at

Morrison; while we, seeing Denver sixteen miles off, started
for it at headlong speed. We reached the Grand Central
Hotel in two hours and a quarter. Then we went to Paul
Graham's, where our clean clothes were stored, and when
we made our appearance at the hotel, dressed like white men
once more, the clerk failed to recognize us—and he was not to
blame. After having a good supper we went to the post-
office, where we found a bundle of letters and papers which
were eagerly investigated, and the evening spent in reading
up the news. Osborn came in at eight P.M.

JULY 26, THURSDAY.

· We were up early and went to the store to sort out and
pack our various boxes. The morning was spent at this
work, and everything was done by noon. We shipped the
rest of the provisions to Dr. Brackett, at Fairplay, and upon
reaching the hotel found General Kargé with the rest of the
party. We spent the afternoon resting and talking with the
party who had just come in, about their ride across South
Park.

JULY 27, FRIDAY.

The day was spent in packing our things up and getting
them down to the depot, and in making arrangements for our
trip northward on the morrow. We had seen enough of
Denver, and wanted to be off.

JULY 28, SATURDAY.

The General had us all up by five o'clock this morning, and
after breakfast we started for the depot with our horses.
Then from 7 till 12.30 we rode over one of the dryest and
warmest stretches of country that can be imagined. Reaching Cheyenne we fed our horses and exercised them, and
then shipped them for Fort Bridger. We then took possession of our palace-car "Summit" and boarded for one day at
the Interocean Hotel, which treated us very fairly. We had
choice bits of game, etc., which were eaten with much rel-
ish. The train at three P.M. had three companies of infantry
bound eastward to quiet the riots. Cheyenne, it may be noted,
is famous for its gambling-saloons and general dissipation.

JULY 29, SUNDAY.

Most of us went to hear the Presbyterian preacher. He
was an old Princeton man, and of course was very much

interested in us. We strolled around the town and asked some questions. The city is in the open plain, without a shade-tree or irrigating ditch. It is particularly quiet now, we were told, as the Black Hills fever had drawn off the scum of the population. At 1.30 we were off for Fort Bridger, and the ride proved very much more agreeable than the prairie scenery, as some hills were always in sight, although one would never think that he was crossing the Rocky Mountains, even when at Sherman, the summit of the Cheyenne Pass. We passed through a great many snow-sheds during the day, reached Laramie at six, and passed the fort of the same name, about two miles before reaching the town. Laramie seemed to us one of the nicest little plain cities we had yet seen. We took supper further on at Rock Stream station.

JULY 30, MONDAY.

We took an early breakfast at Green River station. The scenery, as we passed on from this point, was all new to us and was very grand. The Green River cañon is a picturesque place from the beauty and extent of its erosion, and we spent the short time we had to see it on the platforms of our car enjoying it. We soon turned out of the eroded valley and went into the " bad land" district This was just as interesting in another way, for everything seemed dead now ; not the slightest signs of life, and the fantastic shapes rounded out from the rock on all sides of us seemed the monuments of one vast burying-ground. We reached Carter's station at 10.30. and found our horses there, pretty stiff and thin. We at once took them out to graze and water. Scott was sick, and we left him in charge of the station-keeper's wife, while we rode over to the Fort. After a ten-mile canter we turned the corner of a butte, and the beautiful valley lay before us. The situation seems to be well chosen ; the Fort is well laid out, and there is an abundant supply of both water and wood in the valley. We arrived at 2 P.M. and went to Rickard's Hotel. General Kargé had us all fixed up as soon as possible, and we set out in a body to pay our respects to the dignitaries of the place. We found General Flint, the commander, and Judge Carter very pleasant and agreeable persons. They both expressed a willingness to do all they could to further our plans. Our reception was exceedingly hospitable. The Fourth Infantry band gave us a serenade in the evening, a compliment which we appreciated very highly. Captain Merriman, Lieutenants Brown, Scott, and Adjutant True called in the evening.

JULY 31, TUESDAY.

Speir, Osborn, and Lynde went to Bridger Butte with Mr. Rickard. Potter went over to the station. Scott still unwell. M. looking up guides, and General Kargé and L. making other arrangements. We all attended the open-air concert in the evening, which we found quite a pleasant feature. There was also an inspection in the morning, but as there were only sixteen soldiers out it was not much of an affair.

AUGUST 1, WEDNESDAY.

The "bone-pickers," as we called the rest of our party, were very busy packing up their own peculiar part of the baggage and getting ready for a start. We found a guide by the name of Taylor for them. They got off about 1 P.M., and left us alone to work for ourselves. We called on Gen. Flint in the evening, and were entertained by his charming family.

AUGUST 2, THURSDAY.

Mr. Hamilton brought us in a guide this morning. We liked his looks very much. He has the very prepossessing name of " Santa Anna Joe" (or Joe Oreda), and is a Mexican of short and muscular build. We three selected our stock of utensils for camp life, and then packed everything with our instruments on two mules, which were loaned us by the commissary department. Finally everything was on board, and L. and Joe started for the St. Louis Mills, twenty miles to the south, while M. remained at the Fort. We now commenced our chain of observations from here to the mountains. The first were made this evening between these two points.

AUGUST 3, FRIDAY.

Observations were also made this morning by both of us, M. starting on after them for the St. Louis Mills. L. and Joe in the meantime amused themselves by going shooting. Came back with some birds and a jack-rabbit. When M. arrived we had our first game dinner. L. and Joe then started on again for Gilbert's Meadow. Went up the valley (which was rather thickly wooded) for three miles to Steel's Mills, then, turning to the left, crossed the West Branch of Smith's Fork, then three miles further on up a hill and through the woods till they came out in a beautiful open meadow, where the camp was pitched in a splendid grove of pine-trees near a little brook in the southwest corner. Observations were made at six, and then, while L. was fixing camp and waiting for M., Joe went out and pretty soon a rifle-shot was heard; then all was quiet, until Joe walked in a little later with a deer's heart and liver. M. came soon after, and we had the liver for supper, which we found delicious. We then built a good fire, and after rolling up in our blankets with our feet towards it, went to sleep lulled by the music of the pines. Occasionally a wolf or an owl gave variety to the dulcet tones.

AUGUST 4, SATURDAY.

We did not sleep much last night on account of the cold. We saw some deer, but they ran away before we got a chance at them. We were paid, however, in a moment for our disappointment, as two elk appeared and came down for water, and we shot both; one got away, as he was only wounded. After breakfast Joe went back to Bridger with some instructions, and left us to take care of the elk and the deer. We sent a quarter of the deer to General Flint by him. Our knives were rather dull, and we spent the morning cutting the two animals up. Then came a difficulty; we could not get the mule within fifty feet of any blood, and when we came near him with bloody hands he would rear and kick like all possessed; so we had hard work to get the deer on him; then we took him back to camp and brought out Joe's old black horse, which stood it better. After quartering our elk we found we could only manage one quarter between us, and even then it was not much fun. We carried this back and then got our dinner. Then we set about arranging our camp. By evening all was completed, and we sat down to a comfortable supper of venison. We saw more deer, but were too tired to go after them.

The day was spent in reading, writing, sleeping, and eating in due proportions. We found our camp a delightful spot, and looked forward with much enjoyment to the few weeks we were to spend here. Just at evening, though, we experienced some trouble from the mosquitoes. We thought we were up too high for these pesky little fellows, but found that they would crawl under the leaves when it was cold, and thus save themselves for another time. We experienced some difficulty in our first attempts at bread-making. We also found that the ink we had with us was copying ink, and very thick at that, so we could not persuade our letters to dry at all.

Joe and M. set out to climb Gilbert's Peak (we had been told we could easily), as it only seemed about six miles off. They saw, when they reached the ridge bordering the East Branch of Smith's Fork, that this valley alone was about a thousand feet deep, eight miles wide, and full of little wooded moraines; they then concluded that the best thing would be to move our camp nearer the mountains. So they looked for another camping place. Going along the ridge they came to the head-waters of the West Branch, and there found a very nice place, where they made observations and returned. L. made corresponding observations at the camp. Joe captured the bread-making secret this evening, succeeding finely.

We began to pack early this morning, and got off at eight o'clock. We followed the ridge again, and reached the chosen place at twelve. We spent the afternoon getting things to rights. We used the elk skin for a bed; found it very nice and soft, but it had the fault of letting us out of the tent at the bottom. For an hour or two the mosquitoes were fearfully thick. Our tent is made by supporting a wagon-cover over a ridge-pole, which was driven into a tree at one end and supported by a stake driven into the ground at the other. A blanket is drawn across one end and the other left open toward the camp-fire. Numerous pegs made of hard wood were driven into the softer pine-trees and served as supports for our various articles. We have very little trouble with our camp-fire. A few feet behind us there is an unlimited supply of old pine stumps, which can be cut away with a few skilful strokes of an axe, and when one is set on the camp-fire, after supper is over, it gives a splendid heat and light all night.

AUGUST 8, WEDNESDAY.

Last night it was fearfully cold; our little stream being covered with a two-inch coat of ice. We devoted the whole morning to laying out a base-line. Having chosen two open places on the long ridge which we had come over as the position for it, we levelled our stakes by the transit, and measured both by steel tape and levelling-rod (the transit having stadia wires for this purpose). We were pleased with the results, as we came within $\frac{3}{10}$ of an inch. We were forced to cut a line through a small strip of timber which separated our two open plateaus. These places were almost on a level, and were not only easily seen from all the mountains of the valley, but were beautifully marked out for us by the ring of open ground around each station. They were so plain that missing them was impossible. In the afternoon we fixed our station, No. 3, on the ridge back of the camp, on the other side of the West Branch, and thus closed our first triangle. Held a consultation in the evening, and decided on the plan of going up one side of the valley of the East Branch, and after we had gone as far as we could, to change our camp to the middle of the valley, and come down on the other side. We still mourn the mosquitoes.

AUGUST 9, THURSDAY.

M. and Joe set off with the mules up the ridge. They fixed stations Nos. 4 and 5, measuring both the triangles and the barometric height. They met an Indian, who told them that there were plenty of Shoshones, Snakes, and Bannocks in this part of the country, hunting and setting the woods on fire. L. started the meteorological journal to-day, making observations every fifteen minutes. We had broiled grouse for supper.

AUGUST 10, FRIDAY.

The same programme was followed to-day. M. and Joe set off up the ridge and fixed Nos. 6 and 7. L. continued the meteorological record every fifteen minutes during the day. We killed a two-year old buck, and took him home to camp. This addition to our store of meat proved very acceptable, as we had brought but a small portion of our bountiful supply from the other camp. The night was warm and the mosquitoes stayed late.

AUGUST 11, SATURDAY.

L. and Joe started across the valley to fix two stations on the other side (Nos. 9 and 10), in which they succeeded, but found the road very rough. They made observations for a new camp in the middle of the valley. M. made corresponding observations in the camp. The rainy season has commenced, and we expect a drenching every day. L. found plenty of rain and snow on the ridge while working there.

AUGUST 12, SUNDAY.

We turned out slowly this A.M., and as we thought that all our Indian friends had left us to go across the range, we left our camp to take care of itself, and started for Steel's Mills, a point half-way between the St. Louis Mills and Gilbert's Meadow. This was on the West Branch, so we followed it down. We found an old Indian trail and followed it till we came to a good road, which brought us, after a three miles' ride, to the Mill. We found Dick Steel there with about eight or ten men (as many more having gone off hunting and fishing). The scenes of the day were quite varied. We all sat around the camp-fire and heard stories, until the subject of shooting came up, when Joe was hauled out for exhibition. (He is, by the way, an excellent shot.) He wanted to shoot for ten dollars a side for each shot with the "old hunter," who was providing the camp with meat, but the old fellow was not at all anxious for fame, as Joe's powers had been known there for some years. After dinner they divided into two knots, to play poker. We, of course, watched the crowd Joe was in, and saw the fellow walk off with five dollars of their money by supper-time. They were nearly short of food, so we had the usual camp diet—venison, bread, coffee, and stewed dried currants for both meals. Starting at six we got back to our camp at eight, and found everything as we left it.

AUGUST 13, MONDAY.

We moved camp to-day, so we were all up bright and early.
Before we went M. went to No. 1, and we observed the barometers to get the height of our base-line. Then M. and Joe
started for the new camp with the mules and our utensils,
while L. remained in the old camp to make another series of
observations. L. then went over, and the new camp was in
good shape by night-time. We found our old friends, the
mosquitoes, here in large numbers, and with a good prospect
of their staying all night.

AUGUST 14, TUESDAY.

M. and Joe started early for the ridge. Finished Nos. 8,
12, and 13, and joined in their triangles the point of No. 11.
(This is a very precipitous point in the centre of the valley
which we called Santa Anna Mount.) The scenery was remarkably fine, overlooking the upper basin of the East Branch
on one side and the valley of Black's Fork on the other.
They were troubled by the cold, the snow, and rain, and the
electricity, which affected the transit, sung and hissed in the
monument, gave repeated and severe shocks to the operator.
and at one time ran in a stream from the barometer to the
ground. L. was found alive, but battling with the mosquitoes ; all that was to be seen of his face or body was a ring
around the eyes, left for observing purposes. M. hurt his
knee badly in the timber.

AUGUST 15, WEDNESDAY.

L. and Joe started off this morning, while M. did up some
camp chores, and made corresponding observations. Points
14, 15, 16, and 17 were fixed, establishing some very impor-
tant stations in the valley—one on the transverse axis of the
range. They were also troubled by electricity, and kept from
work for some time by it.

AUGUST 16, THURSDAY.

M. and Joe started up the eastern ridge to point No. 10;
but on arriving there found it impossible to use the transit
on account of the electricity, so putting it up in its case they
went to the next peak in line and built a monument, and were
then forced to retire in a furious hail-storm. To-day, we may
say here by way of parenthesis, our troubles began. We
found this morning one of our mules (the best one at that)

6

had put his foreleg through the lariat-rope around his neck and this had thrown him down. Of course, after thumping around among the rocks all night, he was strangled and nearly dead when we found him in the morning. We nursed him during the day, and at one time thought he might survive, but in vain, when night came .
con gave out, and our meat : :
there were plenty of trout, ai '.

L. and Joe tried hard to ga 	 i;
had to give it up; but they g'   ;
of a deer, only part of which |. _   '. ᴗ.ᴜₙ ᴜᴏᴜᴜᴇ, as they were already loaded with instruments. It rained almost incessantly. We were very much disheartened, and talked seriously of doing other work in the way of returning and joining Fort Bridger to our system. The experiences of the morning with electricity were rather more alarming than usual. No. 18 (which we never included in our triangulation) was shattered to pieces before our eyes, when we were scarcely one hundred paces distant.

AUGUST 18, SATURDAY.

M. and Joe started once more for the same ridge, but were again disappointed by the storms. They built a fire to keep warm by, under one of the ledges, and when they saw that the rain was likely to continue, went off hunting. They succeeded in getting another deer, and brought home the best parts of it. It commenced to rain torrents just after supper, and rained all night.

AUGUST 19, SUNDAY.

We all went to the mill to-day, passing through our last camp about sunrise. We found a bear had been there after our meat that was left, but the fellow was gone, much to Joe's and our disgust. We arrived at 10.30 A.M., just in time to see Steel drive up with the wagon. There were lots of papers and a bundle of letters for both of us, and we enjoyed a regular treat, as well as distributing them to the various mill hands. Old Jimmie, the cook, made a very good dinner, with the rare additions of hot biscuit and pie. After this, the hands went after Joe to get some of their money back, but only got seventy-five cents. We held a council of war, and L. was to go to the Fort to-morrow to see General Kargé and make further arrangements. M. and Joe returned to camp.

AUGUST 20, MONDAY.

L. spent the day riding to the Fort, experiencing a very
. . . . . the road; and, being the only promi-
. . . prairie, was not enchanted with the
spent the day hunting and fishing
. . . fish), and watching the Indians, of
number camped near us.

AUGUST 21, TUESDAY.

L. spent the day at the Fort. He had to content himself with writing messages to General Kargé, as no one was in the Fort from our other party. M. and Joe spent the day fishing, and made themselves sick by eating too many trout. In the afternoon L. left his own horse at the Fort, and took a black one and a little mule from the herd-house, with which he reached Steel's mill safely, and staid there over night.

AUGUST 22, WEDNESDAY.

L. found, on arising, that both horse and mule had disappeared, and started in pursuit of them. He found the mule, but chased him six or seven miles before he caught him. The horse was afterward found near the camp, but would not be caught. L. started back on the mule in the afternoon; he reached camp at four o'clock, and found M. and Joe in a quiescent state—the effect of yesterday's troubles.

AUGUST 23, THURSDAY.

Joe went down to the mill for the horse this morning. L. went up around No. 11 to make some sketches and measurements, and M. staid in camp. Some physical measurements were made of the inclination of strata, the height of timber line was ascertained, and a sketch of the valley as seen from a ledge high up on the precipitous side of No. 11, from which a beautiful view was obtained, far out into the valley of Henry's Fork, and over the bad lands. Joe came back at 6 P.M., with no horse. We decided then and there that we would not be likely to get another horse up to that camp, and that we had better move to our last camp.

We were up early, and packed up everything; then M. and Joe went on ahead with the books and instruments to the other camp. L. staid behind to make observations, and then walked over to the camp. Joe went on with the mules to Steel's mill. M. and L. went off on a partly hunting and exploring trip in the afternoon. Joe came back in the evening, with an old black horse for L. to ride.

L. went over to Steel's mill, to do some writing while M. and Joe went hunting. We did not like to say good-by to the mountains.

M. and Joe joined L. at the mill, and concerted plans for the return. We received a bundle of letters and papers as usual.

M. and Joe started for the herd-house in the morning early, and observations were made between the two barometers. After dinner, they went from there to the Fort, where they made observations also. L. followed them, stopping at the same places until the time appointed for observations arrived. He also filled up the intervening time sketching in the profiles of the adjoining country for a map to show the relation between the Fort and the mountains. He arrived there in the evening. We found everybody well and glad to see us.

The next few days were spent in the pleasant company of those at the Fort, and were enjoyed very much. The Palæontologists arrived slowly, one after another, bringing a rich lot of fossils. We packed them all safely; and then, bidding good-by to our friends at the Fort, we joined the rest of the party, under Dr. Brackett, who were now at Cheyenne, awaiting our arrival. From this point the trip home was very similar to the outward one, with the exception that we passed over the Union Pacific to Omaha, and then to Chicago, where we spent Sunday. The various members of our party were left along the road at the nearest points to their homes, to which all returned with great pleasure, after an experience which will long be remembered by every one who had the honor and privilege to belong to the expedition.

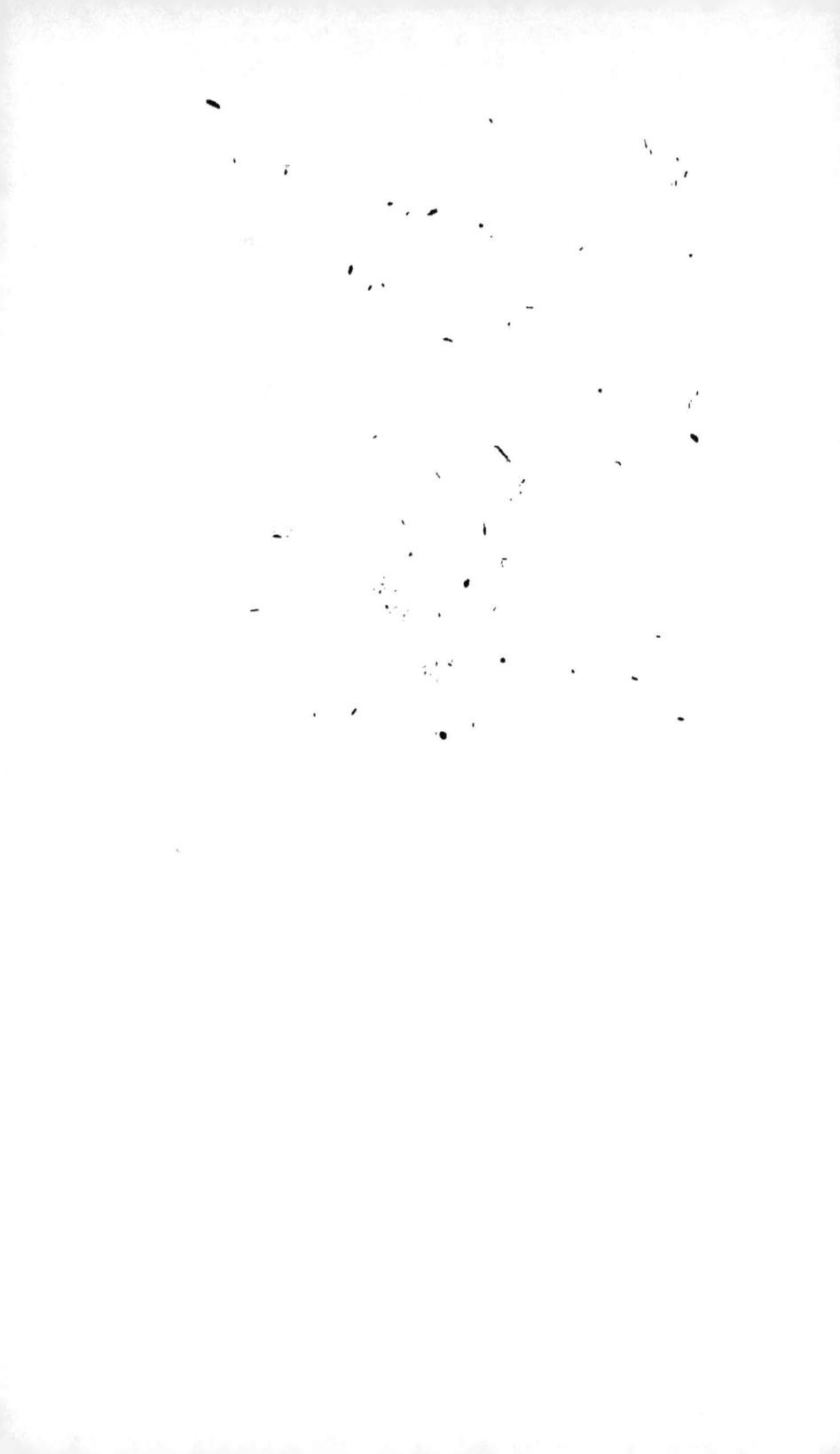

www.ingramcontent.com/pod-product-compliance
Lightning Source LLC
Chambersburg PA
CBHW021942190326

41519CB00009B/1107